CONSTANT TOUCH

A GLOBAL HISTORY OF THE MOBILE PHONE

JON AGAR

Series Editor: Jon Turney

REVOLUTIONS IN SCIENCE
Published by Icon Books UK

Published in the UK in 2003
by Icon Books Ltd., Grange Road,
Duxford, Cambridge CB2 4QF
E-mail: info@iconbooks.co.uk
www.iconbooks.co.uk

Published in Australia in 2003
by Allen & Unwin Pty. Ltd.,
PO Box 8500,
83 Alexander Street,
Crows Nest, NSW 2065

Sold in the UK, Europe, South Africa
and Asia by Faber and Faber Ltd.,
3 Queen Square, London WC1N 3AU
or their agents

Distributed in Canada by
Penguin Books Canada,
10 Alcorn Avenue, Suite 300,
Toronto, Ontario M4V 3B2

Distributed in the UK, Europe,
South Africa and Asia by
Macmillan Distribution Ltd.,
Houndmills, Basingstoke RG21 6XS

ISBN 1 84046 419 4

Text copyright © 2003 Jon Agar

The author has asserted his moral rights.

Series editor: Jon Turney

Originating editor: Simon Flynn

Typesetting by Hands Fotoset

Printed and bound in the UK by
Mackays of Chatham plc

Contents

Acknowledgements v

Part I – World in Bits 1
1 What's in a Phone? 3
2 Save the Ether 16

**Part II – Different Countries, Different Paths
to Mobility** 29
3 Born in the USA 31
4 The Nordic Way 44
5 Europe Before GSM: La Donna è Mobile,
 Die Männer Nicht! 52
6 GSM: European Union 56
7 Digital America Divided 67
8 Mob Rule: Competition and Class
 in the UK 70
9 Decommunisation = Capitalist Power
 + Cellularisation 90
10 Japanese Garden 94

Part III – Mobile Cultures 103
11 Txt Msgs + TxtPower 105
12 Two Organisations in the Congo 111
13 The Nokia Way – to the Finland Base
 Station! 113

14 Mobile Phones as a Threat to Health 122
15 Cars, Phones and Crime 129
16 Phones on Film 143

Part IV – Reassembling the Mobile as a Global System 151
17 The Globe is Made by Standards 153
18 Perpetuum Mobile? 161

Txt Msgs: A Transcript 169
Bibliography 170

ACKNOWLEDGEMENTS

Many thanks to: Cath Skinner, Jon Turney and Simon Flynn for their comments on an early draft of *Constant Touch*; the staff at the British Library and the City Business Library for their help in locating documents; Mark Tindley for some excellent impromptu research on health matters (and to the rest of the Thursday night football crowd for their 'encouragement' to finish); Rob Agar for the chip-munching story; Rico Azicate for pointers towards the Philippines; Jack Stilgoe for advice on health; Ray Martin at BT Archives for help with the illustrations; Sandra Stafford for her copy-editing; and Brian Balmer for a good anecdote.

DEDICATION

To Kathryn and Hal

PART I

WORLD IN BITS

· CHAPTER 1 ·

WHAT'S IN A PHONE?

You can tell what a culture values by what it has in its bags and pockets. Keys, combs and money tell us that property, personal appearance or trade matter. But when the object is expensive, a more significant investment has been made. In our day, the mobile or cell phone is just such an object. But what of the past? In the seventeenth century, the pocket watch was a rarity, so much so that only the best horological collections of today can boast an example. But if, in the following century, you had entered a bustling London coffee shop or Parisian salon, then you might well have spied a pocket watch among the breeches and frock-coats. The personal watch was baroque high technology, a compact complex device that only the most skilful artisans could design and build. Their proud owners bought not only into the ability to tell the time, but also into particular values: telling the time mattered to the entrepreneurs and factory owners who were busy. As commercial and industrial economies began to roar, busy-ness conveyed business – and its symbol was the pocket watch.

At first sight it might seem as if owning a pocket watch gave freedom from the town clock and the church bell, making the individual independent of political and religious authorities. Certainly, possession granted the owner powers over the watchless – power to say when the working day might begin and finish, for example. While it might have felt like liberation from tradition, the owner was caught anew in a more modern rationality, for, despite the fact that the pocket watch gave the owner personal access to exact time, accuracy depended on being part of a *system*. If the owner was unwilling personally to make regular astronomical observations, the pocket watch would still have to be reset every now and then from the town clock.

With the establishment of time zones, the system within which a pocket watch displayed the 'right' time spread over the entire globe. Seven o'clock in the morning in New York was *exactly* twelve noon in London, which was *exactly* eight o'clock in the evening in Shanghai. What is more, the owner of a pocket watch could travel all day – could be mobile – and still always know the time. Such certainty was possible only because an immense amount of effort had put an infrastructure in place and agreements had been hammered out about how the system should work. Only in societies where time meant money would this effort have been worth it.

Pocket watches provide the closest historical parallel to the remarkable rise of the mobile cellular phone in our own times. They started, for example, as expensive status symbols, but by the twentieth century most people in the West possessed one. When cellular phones were first marketed, they cost the equivalent of a small car – and you needed a car to transport them since they were so bulky. But in 2002, global subscriptions to cellular phone services passed one billion. In countries such as Iceland, Finland, Italy and the UK, more than three-quarters of the population owned a phone, with other countries in western Europe, the Americas and the Pacific Rim not far behind.

Like the clock, the phone had made the leap from being a technology of the home or street to being a much rarer creature indeed: something carried everywhere, on the person, by anybody. So, if pocket watches resonated to the rhythm of industrial capitalism, what values do the ringtones of the mobile phone signify? What is it about humanity in the twenty-first-century world that has created a desire to be in constant touch? To answer this question, I began with a rather drastic step.

In this world of weightless information, there is nothing quite so satisfying as taking a hammer to a piece of technology. My old mobile phone, a Siemens S8, was a solid enough device and it took some effort to take it apart. But in the interests of research I wanted to know what was inside. Now the various parts lie in front of me. The battery came away first. It is about the size of a large postage stamp, and as heavy as a paperback book. But I can at least lift it, and this would not have been true of the early years. Let's take an example of mobile radio communication from far back.

The entrepreneurial ex-pat Italian inventor Guglielmo Marconi (see opposite) had hawked his new technology of wireless telegraphy – radio communication by Morse code – around London in 1900, but he focused his efforts on one main customer: the mighty Royal Navy. His pitch was simple. The Admiralty had invested many thousands of pounds in battleships that became incommunicative and blind as soon as fog descended. Wireless telegraphy provided new mechanical senses: to warn of maritime dangers and to organise the fleet. The Sea Lords were convinced and Marconi sealed the deal. The Navy installed thirty-two wireless sets aboard ships. Sixteen years later, during the Battle of Jutland, the purchase would prove a wise one: listening stations detected the German fleet by

Guglielmo Marconi. (By courtesy of BT Archives.)

picking up unusually heavy radio traffic, and the same radio technology enabled the British Grand Fleet to steam across the foggy North Sea to the Danish coast to engage its enemy in the Skagerrak.

Marconi wanted to sell wireless to the Admiralty because a battleship had the size and power to carry it. Radio transmission in the 1900s had been achieved by creating bursts of sparks generated by immense electrical voltages. (The same principle is behind the crackling interference caused by lightning.) Giant voltages meant heavy batteries. The first mobile radio was restricted to behemoths. But a feature of the history of electrical technology has been continuous miniaturisation of components. Even before the First World War, the Swedish electrical engineer Lars Magnus Ericsson had demonstrated new possibilities for mobile communication.

The young Ericsson had trained as a smith and as a mining and railway engineer before becoming an apprentice under the telegraph-maker A. H. Öller in the early 1870s. He then studied abroad in Switzerland and Germany before setting up his own company in Stockholm in 1876, first to manufacture and repair telegraph apparatus, and later, following Alexander Graham Bell's invention, telephones. Ericsson's business boomed. However, he seems to have wearied of the commercial life early in the new century and, backed by a healthy bank balance, retired to a comfortable life as a farmer. But in 1910, and in the spirit of tinkering, Ericsson built a telephone into his wife Hilda's car,

An Ericsson table telephone, c. 1900. (By courtesy of BT Archives.)

the vehicle connected by wires and poles to the overhead telephone lines that had sprung up even in rural Sweden. Enough power for a telephone could be generated by cranking a handle and, while Ericsson's mobile telephone was in a sense just a toy, it did work.

At one level, the story of the retired Swedish engineer-turned-farmer is trivial: no great industry of car-carried mobile telephones was founded on the experiment. But in many other ways it was significant. Ericsson's company, after many twists and turns, would supply much of the infrastructure for the cellular phone systems built in the late twentieth century (more of which later). Second, the experiment happened in Sweden, and the

Nordic countries have a remarkable prominence in the history of the mobile phone that will need to be explained. Finally, Ericsson's charabanc showed that the technologies of communication could be fitted in an automobile, the first instance in a long and profound association between two technologies of mobility that have shaped our modern world.

Marconi's weighty wireless had to be carried by battleship. Early practical mobile phones were carried by cars, since there was room in the trunk for the bulky equipment, as well as a car battery. One of the most important factors allowing phones to be carried in pockets and bags has been remarkable advances in battery technology. As batteries have become more powerful, so they have also become smaller. Partly because improvements in battery design have been incremental, their role in technological change is often underestimated.

The great Prussian physicist Walther Hermann Nernst, who later articulated the Third Law of Thermodynamics, had experimented in Göttingen in 1899 with nickel as a means of converting chemical energy into electrical energy. Built a century later, my disintegrated phone has a Ni-MH – Nickel Metal Hydride – battery. In one sense, this battery is recognisably similar to Nernst's, but in another it is transformed: it is many, many times lighter and more efficient.

Step by step, nickel batteries have become better. Continuous experimentation with other metals has revealed slight but significant improvements, so that, for example, the early twentieth-first-century choice for mass-produced energy packs is between nickel and lithium-based techniques. Gradual change can eventually trigger a profound revolution. Once batteries became powerful *and* portable, a Rubicon was crossed. Uncelebrated improvements in batteries, put into laptops, camcorders and cell phones, triggered our mobile world.

A similar story can be told of the other bits and pieces in front of me. The liquid crystal display (LCD) – the grey panel on which I read my incoming call numbers or Short Message Service (SMS) messages – is now commonplace in consumer electronics. The contradictory properties of liquid crystals – fluids that can paradoxically retain structure – had been noted in the nineteenth century by the Austrian botanist Friedrich Reinitzer. He had noticed that the organic solid cholesteryl benzoate seemed to have two melting points; between the lower and higher temperatures the liquid behaved oddly.

But it was not until the 1960s that industrial laboratories, such as RCA's in the United States of America, began to find applications exploiting this behaviour. Again, incremental development followed. LCDs don't produce light; they reflect light, which potentially saves energy, so changes in one component (displays) interacted with another (batteries). Much effort was needed to turn this advantage into a practical one. However, by the 1970s LCDs appeared in calculators and digital watches, replacing the red glow of light-emitting diodes (LEDs).

LCDs are not essential ingredients of a cell phone. We could keep in constant touch with a simple assemblage of the other bits and pieces found in the wreckage of my phone: aerials, microphones, loud-speakers and electronic circuitry. But LCDs are part of what makes a mobile phone more than a mere instrument of communication. We don't just talk. Without the LCD, the extra aspects of the mobile phone – the games, the address books, the text-messaging – all the features that contribute to a rich mobile culture, involving manipulation of data as well as transmission of the voice, would not be possible.

If I had superhuman strength I could hammer my phone into constituent atoms. A new global politics can be found among the dust. Mobile phones use components that can depend on quite rare materials. For example, within every phone there are ten to twenty components called capa-citors, which store electrical charges. Since the

Second World War, the best capacitors have been made using thin films of a metal called tantalum. On the commodities market in the early 1990s, capacitor-grade tantalum could usually be bought for US$30 a pound, sourced from locations such as the Sons of Gwalia mines at Greenbushes and Wodgina in Western Australia (the world's best source of the element). But in the last years of the twentieth century, as more and more people bought mobile phones, the demand for tantalum shot up, and the price per pound consequently rose to nearly US$300 in 2000.

Tantalum, in the form of columbite-tantalite ('coltan' for short), can also be found in the anarchic north-east regions of the Democratic Republic of Congo, where more than 10,000 civilians have died and 200,000 have been displaced since June 1999 in a civil war, fought partly over strategic mineral rights, between supporters of the deceased despot Laurent Kabila and Ugandan and Rwandan rebels. As the price of tantalum increased, the civil war intensified, funded by the profits of coltan export. However, the mobile phone manufacturers are distanced from the conflict. Firms such as Nokia, Ericsson and Motorola buy capacitors from separate manufacturers, which, in turn, buy raw material from intermediaries. On each exchange, the source of tantalum becomes more deniable. 'All you can do is ask, and if they say "No", we believe it', said Outi Mikkonen, communications manager for environmental affairs at Nokia, recently of her firm's suppliers. On the other hand, export of tantalum from Uganda and Rwanda has multiplied twenty-fold in the period of civil war, and the element is going somewhere.

To build a single cell phone requires material resources from across the globe. The tantalum in the capacitors might come from Australia or the Congo. The nickel in my battery probably originated from a mine in Chile. The microprocessor chips and circuitry maybe came from North America. The plastic casing and the liquid in the LCD were manufactured from petroleum products from the Gulf, Texas, Russia or the North Sea, and moulded into shape in Taiwan. The collected components would have been assembled in factories dotted around the world. While the work might be co-ordinated from a corporate headquarters – Ericsson's base is in Sweden, Nokia's is in Finland, Siemens' is in Germany, Alcatel's is in France, Samsung's is in Korea, Motorola's is in the USA, and Sony's, Toshiba's and Matsushita's are in Japan – the finished phone could have come from secondary manufacturers in many other countries.

The phone might be an international conglomerate, but it was put together in different ways in different countries, and shortly we will see how the cellular phone was imagined in different ways according to national context. I will return later to consider what the mobile tells us about our culture, which has adopted it so readily. I will ask how the mobile cell phone fits with changing social structures, why it has become the focus of new types of crime, and what it can signify when it appears in cultural products such as television programmes and movies. For material components alone do not add up to a working cell phone. Indeed, it was the scarcity of a non-material resource that prompted the idea of the cellular phone in the first place.

Chapter 2

Save the Ether

When Lars Magnus Ericsson was driving through the Swedish countryside, he still had to stop his car and wire his car-bound telephone to the overhead lines. If he had pressed his foot on the accelerator, the wire would have whipped out, wrecking the apparatus. It was not a mobile phone in our current sense of the word. Until the last decades of the twentieth century, most telephones were like this. To use them, you had to stand still, because you were physically connected by inelastic copper wire to the national system. A few privileged people – members of the armed forces, engineers, ship's captains – could command the use of a true wireless phone, connecting to the land-locked national system through radio. The reason it was a privilege was that the radio telephone had to fight for a share of a scarce resource: a place on the radio spectrum.

The first radio transmissions were profligate

beasts. Take Marconi's again. Such radio waves generated by a spark would crackle across many frequencies on the spectrum, interfering and swamping other attempts at communication. This problem meant that early radio users had a choice: either find some way of regulating use so that interference was limited, or take a chance with a chaotic Babel of cross-talk. The route to regulation was taken (although not in all parts of the world; for many decades, Italian radio was the liveliest in the world). But even when radio circuits became more tuneable, so that radio transmissions could be targeted within smaller bands of frequencies, there was never enough spectrum to go around.

Governments seized the right to regulate the radio spectrum. In the USA, the authority was the Federal Communications Commission (FCC). In Britain, it was the Home Office and the Post Office, and so on. But since radio waves are no respectors of national boundaries, national governments had to concede that international regulation had to take place too. Here there was a precedent. In the mid-nineteenth century, the question of how to organise global telegraph communications had prompted one of the first truly international organisations to be set up: the International Telegraph Union (ITU), with headquarters in the neutral Swiss city of Geneva. With the arrival of the fixed-wire telephone in the late nineteenth century, it made sense to extend the ITU's powers over the new technology. Likewise, the ITU was on hand to provide an organisational structure to regulate international frequency allocation for radio.

Every few years, giant international conferences would decide, given existing and predicted use of radio, which services should be allocated a small slice of rare spectrum. These highly technical meetings reflected the world as we know it: engineers and bureaucrats sought to balance demands for new lifestyle electronics such as music radio stations with the commercial necessities of reliable navigation aids, and with the conflicting military imperatives of the technological infrastructures of World War and Cold War armed forces.

In these decades, a phone that worked by radio was a simple enough proposition, but was impossible to imagine as a truly everyday and popular device, since there was no way to squeeze its demands into an overcrowded spectrum already dominated by the powerful commercial and military interests of the twentieth century. Each radio would have to work on a separate frequency from its neighbours, otherwise calls would be interfered with, confused, or worse, eavesdropped. So the radio telephone was restricted to a privileged handful.

But in 1947, engineers at Bell Laboratories proposed a radically new means of imagining mobile radio. It was already shaping up to be a vintage year for Bell Labs. Pump-primed by massive expenditure to develop electronics during the Second World War, the peacetime years saw a string of lucrative discoveries. William Shockley, John Bardeen and Walter Brattain had devised the transistor, the electronic component announced to the public in 1948 that would sweep away bulky valves and lead to the revolutionary lightweight electronics of the second half of the twentieth century. Down a few corridors from the transistor pioneers, D. H. Ring, assisted by W. R. Young, put pen to paper, and the result was a description of the 'cellular' idea. It was a means of saving spectrum.

The Cellular Idea

Ring had written down the principles on which your mobile phone works. Imagine a map of New York City and imagine a clear plastic sheet, ruled with a grid of hexagons, placed over it. Now, imagine a car, equipped with a radio telephone, driving through the streets, passing from hexagon to hexagon (see diagram on page 20).

Ring's idea was as follows. If each hexagon, too, had a radio transmitter and receiver, then the radio

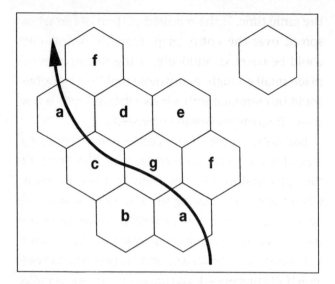

Re-using frequencies creates a pattern of 'cells'.

telephone in the car could correspond with this 'base station'. The trick, said Ring, was to allocate, say, seven frequencies to a pattern of seven hexagons (**a–g**), and repeat this pattern across the map. The driver started by speaking on frequency **a** in the first hexagon, then with **g**, then **c**, and then back to **a** again. If the first and last hexagon were far enough apart so that the two did not interfere (and this was possible, especially if low-powered transmissions were used), then a radio conversation could take place without interference, so long as no one else was in your small hexagon, on your frequency, at the same time. If the repeated pattern of hexagons spread over the entire map, then the whole city could be covered. Suddenly, if the hexagons were made small enough, many more mobile telephones could be crammed into a busy city, and only a few scarce frequencies would be needed.

But notice some of the consequences of Ring's idea, if it was to be put into practice. The driver of the car certainly did not want to have to know when the car was passing from one hexagon to the next. Indeed, when you are walking down the high street, you are passing from cell to cell, yet this 'call handoff' or 'call handover' between cells goes mercifully unnoticed. To conceal the handover, the system needed to be able to spot when the mobile was leaving one hexagon, to find the next hexagon, and to handover the call. In turn, this meant that the individual phone had to be identifiable, so, somewhere in the system, a database needed to be at hand containing information about where the phone was, where it was heading and who was using it (so the call could be billed). That database had to be fast, so it had to be automatic and electronic.

Furthermore, say the driver of the car was in fact talking to his best friend, who was in a phone box in Philadelphia 100 miles away. As the car left and entered each new invisible hexagon, the conversation would be seamlessly pieced together at some central Mobile Switching Centre (MSC), the heart of any cellular network, then fed into the system of old land lines and exchanges, so that finally the distant driver's voice could be heard in the phone box in Philly. For the cellular idea to work, a whole *fixed* infrastructure needed to be in place: base stations, a MSC constantly interrogating a database of personal and geographical information, and connections to the old Public Switched Telephone Network (PSTN). Just as the pocket watch required fixed institutions of agreed protocols and time standards in order that time could be told on the move, so a massive fixed infrastructure of wires, switches and agreements needed to be in place for mobile conversation. Mobility, strangely, depends on fixtures.

Ring had described the cellular idea in 1947, but it went unpublished and, for nearly two decades, gathered dust. Why was this? The reasons were partly technical and partly social.

In technical terms, cellular phones would work best with higher frequencies, where transmissions could be limited to smaller hexagons and where the spectrum was relatively uncongested. But radio engineers had been slow at gathering the expertise to work at higher and higher frequencies through the twentieth century. To find the radio stations on

1920s Bakelite sets, the listener had to tune to up to a few hundreds of thousands of Hertz (Hz, or cycles of radio waves per second, a measure of frequency).

During the Second World War, the demand for better radios and the development of new technologies such as radar had ramped the usable frequencies up to many millions of Hz (Megahertz, or MHz). The first cellular phones worked in the 800–900 MHz band. In 1947, the techniques to handle such frequencies were cutting-edge science. Furthermore, each time the mobile passed from cell to cell, different transmitting and receiving frequencies were used. In turn, this meant that the mobile had to contain a frequency synthesiser, an expensive piece of circuitry to produce the different frequencies. When it was first developed for the military in the 1960s, it alone would cost as much as a very good car.

It's also the case that the technology of switching in the 1940s was incapable of handling the quick call handovers that the cellular idea demanded. The typical switch of the day was that found in the telephone exchange: the clattering, electromechanical relay that was too slow to implement Ring's cellular concept.

However, the answer is never simply technological, for technologies reflect the political and social world in which they are conjured up. Turning

Ring's blueprint into a working cellular phone system in the 1940s would have demanded the latest electronic techniques and a crash course to develop new ones. More importantly, the personal mobile phone fits in with social values that dominate now but did not then. The social world of the mid-twentieth century was hierarchical, paternalistic and even, in large swathes of the globe, totalitarian. The governing model informing both government and business was large scale and top-down. The telecommunications of the mid-twentieth century reflected – and in turn bolstered – this pattern.

I was brought up in Hitchin, a small town to the north of London. In the 1960s and 1970s it was a nice place to be. We lived in a house on the outskirts, with fields out the back and the rest of the estate out the front. We had installed, like everyone else, a squat, heavy, black dial-up telephone by our front door. It was a Type 300 telephone (see opposite). The design had been copied from an Ericsson phone; the Prince of Wales had seen it at the Stockholm Exhibition in October 1932, and liked it.

The telephone was not our possession; it was the property of a government department, the General Post Office (GPO). Only a GPO engineer could take it to pieces if it was broken. (If I had smashed it up,

The black telephone of my childhood. The Type 300 was introduced in 1938 by the General Post Office, and lasted for decades. (By courtesy of BT Archives.)

like my Siemens S8, I would have been in hot trouble. Arguably, it would be treason, a crime against the state.) If the telephone could not be fixed, then only the GPO could supply a replacement. This replacement would be identical to the last phone. And in ironic memory of the long-dead Henry Ford, it seemed as if the Post Office still insisted you could have any colour, as long as it was black. (Other exotic colours – such as ivory – were supposedly available, but I think they were mythical.) The replacement might take a while to arrive. There was a waiting list. A long one.

Nothing better expresses the difference between then and now than this sentiment: we did not mind. We were happy with the squat, black, unreliable telephone. The personal cellular phone would have been almost impossible to imagine in such circumstances. To say that it did not happen because the technology was not there misses an important part of the story. Technology becomes 'there' only when it fits the wider world. Sometime between my early days, growing up in Hitchin, and now, with the pieces of my old mobile phone in front of me, a revolution has happened. A revolution has swept away the GPO monopoly over telephony, and that revolution in turn was part of something bigger – a global sea-change in both politics and technology. Cellular phones are included in this story, and not merely as flotsam but as part of the tidal wave of change itself.

All countries had monopolies providing land line telephones before the 1980s. Even in the USA (where freedom of commerce and industry was prized), the complex Bell System conglomerate, of which the American Telephone and Telegraph Company (AT&T) and Bell Laboratories were part, acted like a monopoly. (To give a sense of scale and power, before it was broken up in the 1980s it was *many* times the size of today's Microsoft.) So a common factor in the following stories of how the cellular phone was built in different countries was a context initially dominated by monopolistic, often government-owned, hierarchical telecommunications organisations.

Sometimes, these PTTs (Post, Telephone and Telegraph administrations), as they are called, were part of the solution. More often, they were the problem. However, the national stories of how cellular phone systems were assembled are very different. By taking apart my old mobile phone, I've found that many material and non-material components are crucial to its working – not only batteries, aerials, chips and LCDs, but also base stations, databases, telephone wire networks, spectrum space and ideas. Bits have come from Finland, Chile, the USA, Germany, Taiwan, Australia or the Congo. But when they were originally assembled in the past, to make cellular phone systems in different parts of the world, the resulting systems were very different. The makers of the mobile phone made history, but not in worlds of their own making.

PART II

DIFFERENT COUNTRIES, DIFFERENT PATHS TO MOBILITY

· CHAPTER 3 ·

BORN IN THE USA

Our engineers & inventors have harnessed the forces of the earth and skies and the mysteries of nature to make our lives pleasant, swift, safe, and fascinating beyond any previous age. We fly faster, higher, and farther than the birds. On steel rails we rush safely, behind giant horses of metal and fire. Ships large as palaces thrum across our seas. Our roads are alive with self-propelling conveyances so complex the most powerful prince could not have owned one a generation ago; yet in our day there is hardly a man so poor he cannot afford this form of personal mobility.

The above message, buried in a time capsule underneath the New York World's Fair, 1939, was addressed to the human race of the year AD 6939. Putting aside the question of whether the future of the seventh millennium would understand 'princes', 'palaces' and 'horses', the message's confident progressive tone rings clear. But the confidence was forced. Not yet a decade of depression after the great Wall Street Crash, and with President Roosevelt's New Deal (and, indeed, Hitler's Germany and Stalin's Russia) demonstrating what state action could achieve, the corporations of the USA were intent to head off criticism, if not revolution, with a spectacular display of how big business was a friend of democracy. Unlike previous fairs, where comparisons between different countries' cultures had been invited, the 1939 New York World's Fair presented the vision of the big corporations – General Motors, Ford, Chrysler, AT&T, General Electric, Consolidated Edison – and the vision was not of the painful past, but of a bright future (see photograph opposite).

In the 'Democracity' of the central Perisphere, Edison's 'City of Light', and particularly in General Motors' exhibitions 'City of Tomorrow' and 'Futurama', visitors were given an eagle's-eye view of the clean, prosperous, technologically driven near-future. From the corporate laboratories would spring miracles in our lifetimes. There was a cure for cancer, while ladies of 75 had 'perfect skin' and buildings were made of plastic. To the 45 million visitors of the World's Fair, the future was promised to offer personal mobility, not only in the form

The New York World's Fair, 1939, promised a bright future – so long as big business was left unfettered.

of cheap US$200 cars, but also through phones without cords.

When visitors left the Futurama display, they were given a badge. On it was a simple statement:

'I have seen the future'. This was the message of the New York World's Fair in a nutshell: while many of the technologies would have been familiar to voracious readers of *Astounding*, *Amazing* and other pulp sci-fi magazines, the Fair made the technological future look imminent. The message of big business was: stick with us through hard times and very soon this future will be yours. So while handheld personal mobile radios had been a fixture in comics and fantasy literature – Dick Tracy, for example, had a wrist-radio, a watch that could talk – the inclusion of radio handsets in the everyday future of the World's Fair made their reality seem, to Americans, just around the corner.

Indeed, heavy car-bound radio communications had been pioneered in the USA in the 1920s. And, unlike Ericsson's experimental system, these were true mobile radios: there was no need to stop the car. Like the fictional Dick Tracy, the first users were trying to stop crime. With the production of fast cars and good smooth roads, criminals were getting harder to catch. The result was an arms race between organised crime and the police, each in turn adopting faster cars, more fearsome weapons, and, to co-ordinate action, speedier communication. The Detroit police force was first to try the experiment in 1921. A patrolman, alerted by a message, would have to stop the car and call in by wire. But in 1928, a fully voice-based mobile radio system was introduced in the city, and other forces followed.

During the Second World War, radio manufacturers, having cut their teeth on police radio, turned to consider military applications. One such company had been started by Paul V. Galvin in Chicago in 1928. But the name of the company, Galvin Manufacturing Corporation, was soon superseded by that of its chief product, 'Motorola' radios, a tag that evokes perfectly the intimate historical relationship of car and mobile radio. One-way Motorola police radios were installed in the 1930s, and the first two-way radio was provided to the police of Bowling Green, Kentucky, in 1940. By then Motorola was gearing up for wartime production. The 'Handie-Talkie' two-way radio was developed for the US Army Signal Corps that year, followed two years later by the 'Walkie-Talkie'. This backpack radio, designed by Daniel E. Noble, worked by frequency rather than amplitude modulation, thereby reducing weight and size while improving performance. Motorola's 35-pound Walkie-Talkie made mobile radio communication practical in the jungles of West Pacific islands or the farmland of Normandy.

All these American systems were mobile radios, but not mobile telephones: you couldn't use a

Walkie-Talkie to speak to someone in a call box. Partly the reason lay in wartime priorities. The telephone network and radio remained separate until 1945, when the war's end meant that military production dropped off and new commercial projects could be given the go-ahead. But the separation was also enforced by regulation. The Federal Communications Commission (FCC) had to be persuaded to drop its opposition before mobile radio telephones could be launched. Nevertheless, the FCC granted a licence to AT&T and South-western Bell to operate the first basic commercial system, called Mobile Telephone Service, in St Louis, Missouri, beginning in 1946. Soon it spread to twenty-four other cities.

Demand for car telephones was intense. AT&T launched its 'highway service' between New York and Boston in 1947, but in New York itself there were severe problems. Not only was there a waiting list of 2,000 potential customers, but 730 lucky users competed to speak on just twelve radio channels. For two decades radio telephony could barely squeeze on to the radio spectrum. Ironically, the congestion was partly caused by the spectacular growth of private radio. '"Mobiling" has become one of the leading activities in ham radio', wrote Charles Caringella, W6NJV, introducing his *Amateur Radio Mobile Handbook* in 1965. 'This growth is only natural; more time is being spent in the automobile than ever before – commuting to and from work, as well as weekend and vacation trips.'

There was not enough room on the radio spectrum for private radio and mobile telephony. But in Ring's concept of the cellular phone, there existed a way out: by re-using radio frequencies in repeating cells, spectrum space could be saved and more users fitted in. AT&T lobbied the FCC, without success, for a decade from 1958 to 1968. Then, at the same time as the civil rights and counter-cultural movements reached a crescendo, there was a reversal of official attitudes. Room was earmarked in the higher end of the radio frequency spectrum for an experiment in cellular telephony, and the radio and electronics industry was invited to respond. Only Bell Laboratories, the research wing of AT&T, replied by the deadline of December 1971. By 1974, the FCC had decided exactly which frequencies the experiment should use, and, by then, Bell had cracked on with the design and construction of equipment, trying it out at the Cellular Test Bed, built around Newark, New Jersey.

In March 1977, the FCC authorised Illinois Bell, the AT&T operating company for Chicago, to install the first cellular phone system. Ten base stations created the cells for an area covering 2,100 square miles around the windy city. At Chicago's Oak Park, the central switch controlled the base stations and linked the system to the public telephone network. It went live in December 1978. By 1981, 2,000 users – the maximum the system could handle – could each speak live from their cars to Oak Park via the cellular network and from there, by the fixed wires, to any telephone in the USA. Other manufacturers followed. In 1980, a Motorola subsidiary, American Radio Telephone Service, began delayed trials in Baltimore and Washington. (Martin Cooper of Motorola had filed cellular patents as early as 1973.) In North Carolina, a small cellular company called Millicom adapted a phone made by the E. F. Johnson firm, producing the first portable cellular phone – for those with strong arms.

These early systems convinced the FCC that cellular radio was practical, and should be deployed systematically across the USA. But, as we shall see, the sheer scale of the country created problems that would not be encountered in Europe. It was clearly beyond the resources of any company – but one – to be able to install throughout the USA, coast to coast, the base stations, switching centres and marketing operations necessary for a national cellular phone system. The one exception was the giant AT&T, the biggest corporation the world had ever seen.

But by the early 1980s, AT&T was under severe attack, leading eventually to the break up of the Bell

System in 1984. Monolithic corporations were not welcome to apply. Moreover, the USA is made up of an archipelago of cities, with great stretches of rural land in between. The FCC decided it made sense to grant cellular licences by auction on a city-by-city basis, the so-called Metropolitan Statistical Areas (any county of a total population of more than 100,000, and that included at least one town greater than 50,000).

This format invited the Bell regional operators – and, after 1 January 1984, the 'Baby Bells', such as Southwestern Bell, Bell Atlantic, Pacific Telesis (Pactel), Ameritech, Bell South, US West and Nynex – to apply independently. In each area, to encourage competition, *two* cellular licences were up for grabs. One would typically go to the local Bell, while the other went to a new company. As part of the deal, AT&T set up a subsidiary, kept at arm's length to satisfy the regulators, called Advanced Mobile Phone Service (AMPS).

The auction of the ninety largest metropolitan areas provoked hysteria. So many applications had flooded in by the June 1982 deadline that the hasty FCC decided that, after awards to the top thirty cities had been made, the other metropolitan licences would go by lottery. This meant that any chancer could land a ten-year licence, then hawk it around more competent operators.

The story was repeated when licences for the less lucrative remaining metropolitan areas and the Rural Statistical Areas were decided by lottery between 1984 and 1989. The last 180 metropolitan area licences attracted 92,000 applications. The end result was that cellular America was extremely disjointed. Each town or city had a different operator, often very local. New companies such as Chattanooga Cellular Telephone, Fresno Cellular Telephone and Long Beach Cellular Telephone served just their local constituency. Roaming, the ability to use your cell phone in different systems (for example, to go from San Francisco to Los Angeles and make calls in both cities), was made extremely difficult.

The disjointed pattern was slowly reversed as the industry consolidated. Firms with licences were bought or merged. By 1992, the largest operator, McCaw, had bought out LIN Broadcasting's and ninety other licences, to serve a total population of over 65 million people. Two years later, McCaw was bought by AT&T; monopolising forces were creeping back. But by then the FCC's management of the licensing process had created a distinctive national cellular style, a crazy-paving of licences covering the country, and had done so slowly. As Garry A. Garrard concluded in an analysis of this licensing phase:

[The USA had] spent four years awarding its cellular licenses for major markets alone, and seven years in total which, when combined with the initial delay in authorising any cellular service at all, gave other countries the chance to catch up on, and overtake, the technical lead originally provided to the US by AT&T.

We will see shortly what other countries were doing to overtake the American lead. But first let's sum up developments so far. Both the concept and first working examples of cellular telephony emerged in the USA, albeit with decades separating the two. The most important factors shaping developments were the existence of the world's greatest electronics-based company, AT&T, *and* at the same time hostility to its monopolistic tendencies. The result was innovation, but in a disjointed form, with many small cellular companies rather than one large one. There wasn't much competition either, since different companies were restricted to different cities. But there was one standard, called AMPS, after the AT&T subsidiary that gave it its name. This standard dictated how a 'terminal' (the mobile) would communicate with the base stations, and would soon have competitors in the wider world.

If there was a distinct pattern of cellular

infrastructure in the USA, there was also a distinct pattern of American cell phone use. Unlike in Europe or Japan, the owner of a mobile phone was charged for accepting an incoming call. This made owners reluctant to give away their mobile phone numbers to all and sundry, and had the effect of making the mobile phone a device for business or emergencies only rather than for chat. The relatively sophisticated – and expensive – cell phone also had to compete with pagers and beepers, which were already very popular with Americans. Stepping off the plane at Baltimore airport in the 1990s the effect, for me, was obvious to see: while in London the mobile was ubiquitous as the prime means of keeping in everyday conversational touch with friends and colleagues alike, on the other side of the Atlantic it was hard to spot a mobile being used – and if it was, the call was brief and business-like. This difference forms the foundation of a quite distinct mobile culture in the USA compared to Europe or Japan.

Finally, there was also a difference emerging in the material design and styling of American phones. For decades, Motorola had led the way in car phones. In early 1984, the company introduced the first hand-portable cellular phone, the Motorola 8000, although since it weighed only slightly less than a pack of sugar, this black brick-sized device was not easy on the elbow. It was hardly an instant commercial success. Four years later, hand portables made up only 6 per cent of sales. We must return to countries such as Finland and Japan to find enthusiasm for well-designed and colourful handsets. Garrard's reasons for the slow adoption of hand portables in the USA include:

> … the high price compared to that for car phones, the fact that early cellular networks were not designed for handportable use resulting in variable or poor quality reception, and the fact that the entire US way of life revolved around the automobile.

But the Motorola 8000 was also a fulfilment of another corporate promise made five decades previously: the wireless personal telephone that formed part of the gleaming near-future of the New York World's Fair.

CHAPTER 4

THE NORDIC WAY

US business producing innovative technology is a story of the 'dog bites man' variety. It is not news. But within two decades of the launch of cellular telephony in the USA, European bureaucracy had produced a technology that more than matched it. As surprising stories go in the history of technology, the success of this European system, GSM (initially, Groupe Spécial Mobile, then later the Global System for Mobile Communications), was definitely man biting dog.

While so often an obstacle to political or technological harmony, the diversity of Europe also means that a wide range of different initiatives can be supported. The development of cellular phones in Europe, for example, directly stemmed from the unique characteristics of one European area: the Nordic countries of continental Scandinavia (Denmark, Norway and Sweden), plus Finland. Only when the cellular idea had been realised in the north did the system spread, for further political reasons, across Europe.

Take Sweden. In the seventeenth century, Sweden was a great power, dominating Scandinavia, Poland and the Baltic trade into Russia. (Indeed, the 'Rus' had been an ancient name given by the Slavs to the colonising Swedish population.) Its style of monarchical government was absolutist (like most of Europe), bureaucratic (like much of Europe) and relatively efficient (unlike much of Europe). This feature, due to the integration of experts of all kinds into the state, survived even as great Sweden declined. Even today, experts such as engineers and academics can move between university and government with far greater ease than in other countries.

Sweden industrialised late, from the 1870s, without forming massive smoke-stack cities like Manchester in England or Lyon in France. The new industries – pulp, paper, ball bearings, matches (Ivar Krueger – the 'match king'), explosives (Nobel), weapons (Bofors), steel, telephones (Ericsson) – employed scientific experts but not many other people, compared to, say, cotton mills. Instead, the unemployed rural population upped sticks and emigrated: a quarter of men, women and children left Sweden, mostly bound for the USA.

This had several effects. A sympathetic bond was

forged across the Atlantic, which helps to explain why American trends (for example, of efficiency and corporate research) could be adopted back home. Most importantly, however, the declining population made government and the owners of industry very unwilling to upset labour, resulting in a distinctive political style: the building of consensus and the rational arbitration of disputes by experts, social democracy and the welfare state. In this society, unlike England or Germany, technology was not, generally, regarded as a threat – even to jobs – but as something good for everyone. Technology was an equalising force. This aspect of the consensus between industry and individuals was reflected in a passion for industrial design (think IKEA), opinion polls and apparently demo-cratising but ultimately paternalistic technologies such as radio.

Swedish industrialisation was also marked by its stance towards natural resources, in particular the swathes of forest in the north. This Norrland provided the raw materials for the pulp, match and paper industries, but it was not ransacked. Instead, Swedes regarded it as a national resource, and one to be used efficiently in a planned, rational manner. Indeed, until recently, Sweden had a very dispersed and sparse population, encouraging identification with the countryside.

An environmental tinge entered into the collection of ideas and attitudes that Swedes regard as their contribution to the world. Fed by a nostalgia for great power status, by a twentieth-century foreign policy of military neutrality, and by pride in the harmony created by 'third way' social democratic politics, Swedes have been keen that the world should learn from their example. Since the Second World War, Swedes have, for example, been highly active in the United Nations (UN). We will see that this peculiar internationalism shaped technology, too.

Swedes tend to bristle indignantly at the suggestion that they are very similar to Norwegians or Danes. Of course no two individuals are alike. But Scandinavian countries share more cultural and political similarities than differences, however jealously cultivated the latter can be. A shared history helps explain this: the three formed the fifteenth-century Kalmar Union, which, with different twists of fate, could have been the foundation of a United Kingdom of Scandinavia.

In the nineteenth century, under Oscar I of Sweden, an explicit project of Scandinavianism was pursued. Then, in 1864, the project's poverty was revealed when Sweden failed to come to Denmark's aid against a Prussian invasion. Still, many of the characteristics I have ascribed to

47

Sweden can also be applied to Denmark, Norway and to some extent Finland – including social democracy, internationalism and an enthusiasm for technology.

In 1967, the chief engineer at Swedish Telecom Radio, Carl-Gösta Åsdal, suggested that an automated nationwide mobile telephone and paging network should be built, integrated with the land line network. Further studies, supervised at Swedish Telecom's radio laboratories by Ragnar Berglund and Östen Mäkitalo, tested the feasibility of Åsdal's idea. For such an ambitious project, collaborators were needed. In 1969, the Nordic Mobile Telephone (NMT) Group was established, with engineers representing Sweden, Denmark, Norway and Finland, and with a view to developing a cellular phone system. These engineers worked for the state-owned telecommunications businesses, and they therefore carried with them the shared values of Nordic government – in particular, a faith in development by consensus and rational discussion between experts. (These common values would not have been found if, say, the gathering had been between English, French, German and Spanish electrical engineers.)

The NMT standard, defining how parts of the cellular system would interact, was also crucially affected by the Scandinavian relationship between

48

expert and the state: while the expert was integrated into government, and therefore retained influence, a close touch with everyday society was preserved. In the spirit of good industrial design and of technology as an equalising force, customer surveys ensured that the NMT standard matched people's wants and desires. (But notice how different the Scandinavian engineers' attitude was from, say, that of the UK's General Post Office. No insistence on one squat black telephone here!)

The political heritage of Scandinavia can also be found in the management of the radio spectrum. Like a virtual Norrland, the spectrum was seen as a national resource, the use of which should not be restricted but planned for efficiency and the greater good. Already in 1955, for example, the Swedish state-owned telecoms company, Televerket, had launched the first mobile telephone system in Europe, less than a decade after AT&T's equivalent. With a small population spread over a big, forested country, mobile radio found many customers. With high demand and official encouragement, there were already 20,000 mobile phone users in Sweden by 1981, proportionately far higher than elsewhere in Europe, when the NMT cellular system was launched. With such a prepared and fertile ground, it is not surprising that NMT was a success. By 1986 capacity was full, and a second, higher frequency system (called NMT 900 to distinguish it from the lower frequency NMT 450) was introduced.

NMT had demonstrated that it was possible for electrical engineers from different countries to sit down together and write a standard for a transnational cellular phone system. It was possible to 'roam', to use the same phone passing from Oslo to Helsinki, say, because the technical details determining how a mobile terminal would communicate with base stations, or how base stations would link to the switching centre, had been hammered out beforehand. (Compare this ease of roaming with the situation across the Atlantic, where the common standard, AMPS, was much looser, and where operations depended to a greater extent on decisions made by each company. It was impossible to roam freely across the privately owned crazy-paving of cellular America.)

Furthermore, once the NMT standard was defined, the telecoms industry, already closely involved in developments, could freely offer products. Mobile phones for the pioneering Nordic cellular system were provided both by home companies (the Danish Dancall and Storno, Swedish Ericsson and Finnish Mobira, part of Nokia), and by American and Japanese companies. The prosperous Scandinavians could afford the price – the equivalent of the small car that was still needed to carry an early mobile phone. However, the contract for the central switching technology, the heart of any cellular system and therefore a strategically important economic decision, went straight to Ericsson.

· CHAPTER 5 ·

EUROPE BEFORE GSM:
LA DONNA È MOBILE,
DIE MÄNNER NICHT!

By 1987, five years after the launch of the Nordic mobile cellular phones, roughly 2 per cent of the population were subscribers. Cell phones had become a standard tool for truckers, construction workers and maintenance engineers, although a few were being sold for private use, especially for installation in the weekend holiday homes and boats that are a feature of Scandinavian life. This early lead would continue into the twenty-first century.

The public telecoms monopolies in other European countries slowly began to respond to the Nordic experiment and to glimmers of public demand. Several – Spain, Austria, the Netherlands and Belgium – ordered NMT services. Often these services stalled, since the exact NMT frequencies were already in use, which meant that new expensive mobile terminals had to be produced. In Spain, for example, Telefónica had ordered the NMT system from Ericsson even before it had been launched in Sweden, but the pricy terminals attracted few customers.

But the chauvinistic telecom monopolies of the 'big' European countries – Germany, France, Italy and Britain – decided to design their own mobile cellular systems. Each was designed to respond to peculiar national demands. In France, the Direction Générale Télécommunications began an ambitious *grand projet*, Radiocom 2000, which was more like a souped-up national dispatch radio system than a cell phone. It was as distinctly Gallic as Minitel, the network of public information terminals that would later act as a barricade against the Internet-pronged attack of American mass culture.

Contracts for Radiocom 2000 were awarded to another immense state-controlled French combine, the military and aerospace firm Matra. Like many French national projects, when Radiocom 2000 was launched in 1985 it was available only in Paris. Again, prices were high and uptake low. However, it can be too easy to be unfair in retrospect. Mobile analyst Garry Garrard records the telling observation of Philipe Dupuis in 1988: 'If it had not been demonstrated in other countries that mobile communications can become more abundant and cheaper, everyone would be happy.'

Likewise, in Germany, the country in Europe in

which wealth and technical proficiency were highest, a new cellular standard was developed by the telecoms monopoly, Deutsche Bundespost, and the dominant electronics manufacturer, Siemens. Netz-C was launched – comprehensively, to 98 per cent of the West German population – as a commercial service in May 1986. A unique feature of early German mobile cellular phones was a personal identification card, the ancestor of the later SIM card. This bonus helped to bump up the price, so again, compared to cell-happy Scandinavia, sales were disappointing.

Politics shaped the German mobile system in another unique way. In 1990, following the fall of the Berlin Wall and the subsequent reunification of Germany, the appalling state of East German telecommunications had to be confronted if the wider tasks of transforming the economy and rebuilding government were to proceed. The situation needed a demonstration of the power of capitalism. Instead of slowly installing telephone cables, the mobile Netz-C was quickly expanded to cover the East German Länder.

But Netz-C was a poor system. Compared to Italian networks, and especially the pioneering Nordic cell phones, German mobile was a failure. We will see soon how an unsatisfied German market would be crucial to launching a new, world-beating mobile system. However, had we paused in 1991 and surveyed mobile Europe, we would have discovered a pattern that would have confirmed the prejudices of any Euro-sceptic. The picture was of a hotchpotch of national systems, each designed to satisfy national interests (often, indeed, the interests of the public telecoms monopolies rather than French, German, Swiss or Spanish customers), and employing ten incompatible standards. Oddly, the spectacle was most pitiable in the wealthiest, most technophilic countries, France and Germany. The United Kingdom, insular in its own way, was pursuing a detached Thatcherite experiment. We will look at its outcome later in this book. But, next, a bureaucratic miracle.

CHAPTER 6

GSM: EUROPEAN UNION

In a chilly Stockholm, in December 1982, engineers and administrators from eleven European countries gathered to inaugurate GSM. An acronym that initially described the countries involved (Groupe Spécial Mobile) but that later became the Global System for Mobile Communications would record an unlikely feat of world conquest.

The lead had come not from the big European powers. It had come from the Nordic countries (which, in NMT, had a successful trans-national mobile system ready for expansion) and from the Netherlands (which did not).

The delegates were in Stockholm to consider whether a Europe-wide *digital* cellular phone system was technically and, more importantly, politically possible. Older standards such as NMT had been analogue, which is good for the transmission of voice, but little else. Going digital created the opportunity to provide new services, such as data transmission. More importantly, it created the chance to make a pan-European political statement. But despite the example of NMT, surely the chances of overcoming entrenched national telecoms interests were slim?

In February 1987, in the warmer climate of Madeira, the group gathered again to hear the results of prototype tests. Not only had these tests been passed with flying colours, but also the political will had been found to iron out national differences and to choose one pan-European standard. What had happened?

The 'Europe' built on the ruins of the Second World War was the outcome of two opposing tendencies. For every Jean Monnet or Robert Schumann who dreamed of a United States of Europe, there was a General de Gaulle or Margaret Thatcher who was deeply suspicious. It is noticeable, however, that while the sceptics of Europe bolstered their arguments by appeal to grand, even sentimental, *ideas* – of national spirit or self-determination – the federalists' project of building Europe has often progressed by mundane and technocratic appeal to better technological, *material* organisation.

So, for example, following the Monnet-authored Schumann Plan of May 1950, the European project was launched with the establishment of the European Coal and Steel Community by the Treaty of Paris in 1951. In six countries (France, Germany, Italy, Belgium, Luxembourg and the Netherlands), the industries that produced the material structure of Europe – the Europe of reinforced concrete, car chassis, dynamos and steel knives – came under the power of a supra-national Higher Authority. In 1957, with the two Treaties of Rome, one community became three, with the addition of a European Economic Community (EEC) and, in the nuclear field, Euratom – another example of the technological spirit behind European organisation. In 1986, by which time the original six had been joined by the UK, Denmark, Ireland, Greece, Portugal and Spain, the Single European Act moved beyond the earlier target of mere co-ordination of the EEC, adopting the much more ambitious aim of making Europe a single market, bigger than any other in the world.

The project to build one European cellular phone system, based on the GSM standard, would be a major material means of realising the dream. As the institutions of Europe grew more powerful, so they increasingly became the sources of further pushes towards European integration. Of these, the European Commission (EC), created from the smaller staff and management of the three existing communities and headed by a president, was the most powerful.

Evidence for GSM as part of a dream of 'Europe' can be found by wading a little way into European bureaucracy. In particular, a Recommendation issued by the Council of the European Communities of 25 June 1987, called 87/371/EEC, addressed the question of mobile communications directly. The Council is a collection of politicians, but the paperwork it considers is largely generated by the Commission. So 87/371/EEC should be seen as an expression of the interests of the EC. The document is written in legalese, but the reasons for favouring a pan-European mobile system shine through. Broadly, they were twofold.

First, the dream of a single European market would remain just that if a means was not found of reducing national differences and improving communication. When the authors of the document wrote that 'the land-based mobile communications systems currently in use in the Community are largely incompatible and do not allow users on the move in vehicles, boats, trains or on foot throughout the Community, including inland or coastal waters, to reap the benefits of European-wide services and European-wide markets', they were addressing both problems.

GSM offered an exceptional moment for reducing difference: the 'change-over to the second generation cellular digital mobile communications system' provided 'a unique opportunity to establish truly pan-European mobile communications'. 'European users on the move', communicating 'efficiently and economically', would be the basis for a single market. Again, 'Europe', an otherwise rather ghostly entity, would be given substance by building material technological systems.

Second, the eurocrats kept a watchful eye on Europe's main economic competitors, Japan and the USA. Not only would GSM be an instrument of European unification, but also it would provide a lead in the cut-throat but strategically important global marketplace for technological goods. 87/371/EEC states explicitly that:

> *A co-ordinated policy for the introduction of a pan-European cellular digital mobile radio service will make possible the establishment of a European market in mobile and portable terminals which will be capable of creating, by virtue of its size, the necessary development conditions to enable undertakings established in Community countries to maintain and improve their presence on world markets.*

So the EC, the civil service of the European project, had seen in GSM a political tool of immense value: telecommunications – and particularly GSM – would provide the infrastructure of a Europe ready to mount a convincing economic challenge to the US and Japan, and a pan-European telecoms network would encourage organisations to think European. The one drawback was that the EC had not been responsible for encouraging GSM from the outset. Instead, the representative of the old, increasingly unfashionable public telecommunications monopolies, the Conference of European Posts and Telecommunications Administrations (CEPT), had been. Never mind. With bureaucratic *sang-froid*, the Commission claimed credit anyway.

The best illustration of GSM as a politically charged European project is given by the facility to roam. Just as in NMT, roaming (the ability to use the same terminal under different networks) was prioritised, even though it was expensive, because it demonstrated political unity. The complexity added to the technical specifications to allow a mobile user to drive from Lisbon to Leiden gave GSM a new, unwelcome meaning – the 'Great Software Monster'. Commentators called GSM the 'most complicated system built by man since the Tower of Babel'. But the political intention was in stark opposition: in place of the conflicting chaos of incomprehensible tongues, GSM would stand for unity.

In practice, very few users had roamed across Norway, Sweden, Denmark and Finland with the

NMT system. Most users had stayed within a few miles of Oslo, Stockholm, Copenhagen or Helsinki. Again, with GSM, for many years roaming had been an expensive political luxury – the telecommunications equivalent of the agricultural subsidies that were grudgingly paid to keep European peace. Ironically, when roaming did take off in the very late 1990s, it was enthusiastically embraced by partying twenty-somethings in Ibiza as much as by European business executives. The reason can be found in a capacity that had been buried in the GSM specifications as little more than an afterthought: the Short Message Service (SMS). The phenomenon of text messaging is examined later.

While GSM was meant to be launched on 1 July 1991, in fact it wasn't ready. But the bureaucratic miracle had to be witnessed, and a few symbolic conversations were organised: Harri Holkeri, the Governor of the Bank of Finland, phoned the mayor of Helsinki, and discussed the price of Baltic herring. But such fishy anecdotes aside, commercial GSM services did not really start until the following year. Eight countries – Germany, Denmark, Finland, France, the UK, Sweden, Portugal and Italy – began in 1992. They were soon joined by others, so that by 1995 European coverage was nearly complete. Indeed, several countries had more than one GSM operator. Remarkably, GSM then began to be adopted outside of Europe. By 1996, GSM phones could be found in 103 diverse countries, from Australia to Russia, from South Africa to Azerbaijan, and even to the USA. (See photograph below.)

The GSM standard for digital cellular phones was a worldwide success. This shop in Dalston, East London, specialises in GSM phones for African networks.

European bureaucracy had scored an undoubted commercial hit. There were many factors behind the remarkable uptake of GSM. A feature of the history of standards is that success tends to create its own momentum. So, to take two notorious examples, the QWERTY keyboard and the VHS video system were both 'chosen' not because they were technically superior to their rivals (they weren't), but because everyone else was already using them.

GSM was not the only digital cellular standard on offer in the 1990s, but since a significant number of countries had already adopted it – albeit for political reasons – it was a safer choice. The equipment was ready to buy, and experience showed that it worked well enough. But this dynamic is not by itself enough to explain GSM. The European digital standard benefited, bizarrely, from the chaos of what went before. Many different national systems at least produced a variety of technical possibilities from which to pick and choose. NMT provided a basic template onto which extra features – such as SIM cards, from Germany's Netz-C – could be grafted. Furthermore, the relative failure of the first cellular phones in countries such as France, and especially Germany, created a pent-up demand that GSM could meet. In the USA, where customers were satisfied with the analogue standard, there was little demand for digital until spectrum space began to run out. Paradoxically, the USA lost the lead because its first generation of cellular phones was too successful.

More important still was how GSM satisfied both European customers and manufacturers. You may recall that when I smashed up my old GSM phone I found it consisted of many components. In the early 1990s, technical trends, especially miniaturisation, led to a qualitative change in mobile terminal design. Suddenly, mobile phones became small and light enough to routinely carry around. ('Hand portables' had existed before, but they were not cheap and certainly not easily manageable.) There was a leap from car phone to hand phone. Part and parcel of the same process, the new designs attracted new customers, and the mobile became less a business tool and much more an everyday object. The shift marks the start of the mobile as mass culture and individual necessity.

Three manufacturers soon dominated the mobile market. Spurred by competition between each other, Nokia of Finland, Ericsson of Sweden and the American Motorola designed and marketed ever smaller and cheaper phones. These companies, and European firms that supplied other parts of the GSM cellular networks, also benefited by the fortunate outcome of a patents battle. Aware that such a complex system as GSM might rely on patents that may only be discovered, ruinously, at a later date, European planners of GSM insisted in 1988 that manufacturers indemnify themselves against the risk. Although the demand was later watered down, many manufacturers, including the mighty Far East conglomerates and most US firms, decided to sit GSM out. This obscure legal controversy had the effect of reducing competition and massively strengthening the firms amenable to agreement: Nokia, Ericsson and Motorola. (Motorola had European factories.) Select manufacturing interests were therefore happy with GSM. These factors were as important in pushing GSM forward as pan-European political motives had been for starting the ball rolling.

Chapter 7

Digital America divided

Only when too many American analogue cell phone users were cramming into the available spectrum space did attention seriously turn to launching a digital standard that could compete with GSM. In the spirit of free market competition, a number of alternative approaches were proposed.

Imagine you have thirty people at a great party, all on the same cell phone network and all calling their friends. If they all tried to use the same frequency at once, then no one would be able to talk. This was the problem confronted by the tele-communications engineers at the very end of the 1980s – except that the party could be the size of Chicago or New York. There are several solutions. Each party-goer could be allocated a tiny sliver of frequency on which they had sole use. This solution, called Frequency Division Multiple Access (FDMA), works only so far as you can slice up the scarce frequency bands thinly enough without affecting operation. In practice, you soon run into problems. So another solution is not to broadcast all the conversation. For example, take a second of transmission. By devoting the first-thirtieth of a second to the conversation from the first party-goer, the second-thirtieth of a second from that of the second party-goer and so on, a small part of every phone call from the party is transmitted. A trick is played on each listener – they don't notice the gaps! This method, easy by digital, is called Time Division Multiple Access (TDMA), because you are dividing up the transmission into different time slots and letting several callers access it.

A third possibility is much harder to explain. It is best attacked from a different direction. Imagine someone at the party wanted to have a secret, illicit conversation. The caller could borrow a technique from the shadowy world of codebreaking. By mixing up the information that coded the secret conversation with a seemingly random signal and transmitting the mixture together, then even when mixed in with other messages using the time division method, the receiver at the other end – if they know what the random signal was – can unpick the original message. What this method – called Code Division Multiple Access (CDMA) – loses in lack of simplicity, it gains in privacy.

The invention of CDMA had its roots in San

Diego County, where there existed a fruitful interplay between military contractors and top electronic research centres, such as the University of California at San Diego (UCSD). Irwin Jacobs and Andrew J. Viterbi, who had met at UCSD in the 1960s, launched Linkabit (a military communications company) in 1968. In 1985 they sold out, and started again with Qualcomm, through which Jacobs and Viterbi hoped to commercialise another product of the Californian military-industrial complex: CDMA, a concept that had been developed when Linkabit had been asked to develop a satellite modem for the United States Air Force.

Qualcomm had the skills to take CDMA from its military setting and to reap the considerable profits of entering the consumer communications market. (There was also a contract from the Hughes aerospace giant to help things along.) Nevertheless, Qualcomm needed considerable powers of persuasion, since the decision was taken in 1992, on competition grounds, to allow both TDMA- and CDMA-based standards to proceed despite CDMA coming late to the party. The result was that the USA yet again resembled a patchwork – this time of a variety of incompatible digital standards.

CHAPTER 8

MOB RULE: COMPETITION
AND CLASS IN THE UK

'I'm on Westminster Bridge.' Telecom Pearl in 1986.
(By courtesy of BT Archives.)

In the early 1990s, something exceedingly disturbing was happening on trains across Britain. People were talking. Loudly. The anger, generated among unwilling eavesdroppers and aimed at the mobile owners cheerfully declaring that they were 'On the train', was a sure indicator that an invisible social boundary had been transgressed.

In the early nineteenth century, the stage coach had been alive with gossip and chatter, as the novels of Jane Austen or the essays of William Hazlitt record. With the arrival of the steam locomotive, however, the talk stopped. Partly the smooth, speedy – almost unworldly – motion of the railway carriage on iron tracks was conducive to a mind more contemplative of the landscape outside the window than to discussion with fellow passengers. Trains transported the body *and* mind. But a more severe problem lay with those passengers themselves. *Who* were they? Railways were symbols of an industrial age, and in the sprawling industrial city, people became increasingly anonymous. Although the division of carriages into different classes – first, second and third – gave some clues, it remained extremely awkward to strike up a conversation on a 'suitable' note. Rather than commit a social gaffe, travellers on trains in Britain chose silence.

Delicate issues of class had created social protocols

of communication – rules governing when to speak and what could be said, rules that may never have been written down but were more powerful for all that. Against a century and a half of mutually sanctioned quiet ran a device created by a new set of protocols. GSM, at heart, was also a set of rules governing communication. But these were hard and explicit, and individuals – not society – accepted them on the purchase of a cellular phone. It was the individual, not society, that spoke loudly: 'I'm on the train.'

Listeners were annoyed. The older tacit rules had been broken. Not only that, their particular complaint against the owners of mobile phones was of *selfishness*. How dare they disturb everyone else! What makes their conversation so necessary, so important, to justify shattering the collective trance? What makes *them* so special? Indeed, the history of mobile phones in Britain is intimately entwined with social transformations, class transgressions, and competition – not only between technical systems but also between the politics of selfish individuality and the social bonds that tie us.

During the summer of 1954, the Marquis of Donnegall was jealous. The Duke of Edinburgh, he had heard, possessed a telephone built into a car. The Marquis' information was correct. The Duke's Lagonda coupé sports car had a radio telephone with which, via an Admiralty frequency and a Pye relay station up on the Hampstead hills in North London, he could speak directly to Buckingham Palace. He enjoyed this perk of the job, as the breezy *Daily Sketch* told its royal-friendly readership: 'The Duke takes a keen delight in making surprise calls to the Queen ... Sometimes he disguises his voice when speaking to Charles and Anne.' (The newspaper also hinted at fears concerning the combination of royalty and speeding cars – 'He is a skilful driver but some concern was felt that he should use so fast a model' – while adding the reassuring statistic that the Duke, as Lieutenant Philip Mountbatten, held the unofficial record among his fellow naval officers for the 98-mile run from Bath to London. In his 12 hp MG he had covered the distance in one hour forty minutes.)

The Marquis of Donnegall asked the Post Office whether such a radio telephone could connect to the public network and, if so, whether he could also have a set. The Post Office's reply is revealing in what it tells us about British mobile radio telephones in the mid-1950s:

You probably know it is possible for passengers on certain ships to make radio-telephone calls to the United Kingdom public telephone system; and shipping in the Thames, if provided with the

appropriate equipment, can also call on-shore
telephones. There is, however, no arrangement for
private persons to fit radio in their own vessels
or vehicles for communication with the public
telephone network. The prospect of starting such a
service in the United Kingdom in the present state
of technical knowledge in the radio field are nil.
There is no room in the radio frequency spectrum.

Actually, the Duke's car phone *could* have been connected to the public telephone system – experiments had proven this capability – but the Duke had balked at the cost. While it was 'Post Office policy to refuse the connection of radio calls from privately operated services (which include those established by police, fire, and public utility organisations) because of the difficulty of maintaining the necessary standard of transmission', an exception could be made for a person of appropriate social standing. In the mid-1950s, if you were the husband of the Queen, then you could have a mobile telephone connection to the public telephone network. But if you were a mere Marquis, you could go whistle.

Putting aside the finer points of social rank, the Post Office's reply also illustrates the typical sentiments of a public telecoms monopoly. It looked inward, rather than outward. The Post Office was more concerned to preserve the integrity of the network than to be led by – or even concede to – customer demands. Early users of mobile radio in Britain included the travelling car repair services (such as the Automobile Association or the Royal Automobile Club), taxi firms (particularly in the capital, such as RadioCabs (London) Ltd), and industrial companies whose facilities were dispersed widely and in far-flung places (such as Esso Petroleum Ltd). Even banded together to form the Mobile Radio Users' Association, they were powerless against the might of the public telecoms monopoly. In 1954, for example, the mobile users were kicked off their frequencies because the government wanted to create room in the spectrum for the commercial rival to the BBC, ITV.

The Post Office consistently (with few exceptions, not least the case of the Queen's husband) refused to consider connection of mobile radio to the telephone network. By 1968, when, against two decades of Post Office lack of interest, 6,100 private mobile systems were licensed (a total of about 74,000 stations altogether, and a growth rate of 17 per cent per year), the official attitude was still that the integrity of the telephone network was paramount and any interconnection of the noisy, anarchic mobile radio to the state-owned system could barely be countenanced: 'The policy of refusing connection of private mobile systems to the public network has lasted nearly 20 years. From the [official] point of view the argument for maintaining this refusal is as strong as ever.'

Actually, by the mid-1960s the Post Office was beginning, reluctantly, to change its policy on interconnection. An experimental South Lancashire Radiophone Service had begun around Manchester, Liverpool and Preston in 1959. In 1965, an extremely exclusive and expensive service, called System 1, had been launched in the well-to-do Pimlico area of

The London Radiophone Service, later called System 1, was launched by a call from the Postmaster General (the government minister responsible for post and telecommunications) to the TV presenter Richard Dimbleby. The Service was not a cellular system, but allowed callers to connect to the public telephone network via an operator at the Tate Gallery Telephone Exchange and base stations at Kelvedon Hatch (near Brentwood, north-east of London), Bedmond (near King's Langley, north of London), and Beulah Hill (in Croydon, South London). (By courtesy of BT Archives.)

West London. It was used by the chauffeurs of diplomats and company chairmen. The radio set cost £350, the service cost over £7 a quarter year, and calls cost 1s 3d (about 6.5 pence) for three minutes.

Two years later, the emergency services were allowed to connect to the telephone network. Indeed, the mobile telephone had begun to trickle down the British class system. On the eve of the introduction of cellular phones, a privileged 14,000 used the later (non-cellular) System 4 mobile telephones. The terminal alone cost £3,000 and the annual subscription was a quarter of this sum again. Car ownership is a good indicator of status, and most users of System 4 drove a Rolls-Royce, BMW, Mercedes or Range Rover.

A class act all round. Woman in a Rolls-Royce Corniche, with radiophone, 1975. (By courtesy of BT Archives.)

The later story of mobile phones in Britain makes sense only in the context of the political transformations instigated by Margaret Thatcher, who was elected as prime minister in 1979. The early years of her Conservative administration were decidedly shaky. She had ditched the consensus politics of her forebears in favour of rolling back the state and a brutal regime of monetarist economics. Unemployment shot up. Riots flared in London, Liverpool and other cities.

Thatcher's drastic experiment on Britain would almost certainly have ended in failure, if it were not for two factors. First, the unexpected war with Argentina over the Falkland Islands rekindled nationalist emotions that few, in particular the ineffectual Labour party in opposition, suspected still existed. Yet, second, Thatcherism made an even stronger appeal to the purse than to the nationalist heart. In rolling back the state, Margaret Thatcher realised that the privatisation of state-owned resources released money, in the form of stocks and shares, that could be sold cheaply to the voter. A virtuous cycle of greed was set up, in which industry was liberalised and individuals enriched. It was to be an experiment in class and competition.

Telecommunications – in other words, the Post Office – provided an ideal test case. Two years after Thatcher's rise to power, the Telecommunications

Act (1981) was passed. The phone business side of the Post Office was stripped away and renamed British Telecom, and a competitor was authorised – Mercury Communications Ltd, a new face to the old imperial firm of Cable & Wireless. While for the moment British Telecom would remain a public corporation (like the old Post Office), a second Act, in 1984, privatised it. The theory was that with the new exposure to the market, British Telecom would be forced to respond to customer needs. The era of the squat black telephone, with one size fitting all, was over.

But Mercury, competing with British Telecom on the fixed-network telephone service, would provide a poor demonstration of the powers of market capitalism. The leviathan British Telecom remained a *de facto* monopoly over the land lines. Instead, mobile telephony became politically hot: since cellular systems would be introduced from scratch, then British Telecom would be on a level playing field with competitors.

In 1982, the government announced that two analogue cellular licences were up for grabs. As an extra hobble, British Telecom was told that a submission would not be welcome directly, only in the form of a joint bid in a minority partnership. Intensely annoyed, but unwilling to yield a promising area of telecoms business, British Telecom teamed up with the private security firm Securicor. (Securicor had done very well in the Thatcher years, and therefore had the money to risk, but it was also an experienced user of private mobile radio.)

This service, operating under the name 'Cellnet', was guaranteed the first licence. The second licence was won in December 1982 by a consortium led by the unusually market-orientated defence electronics firm, Racal, and inspired by Gerald Whent, then chairman of Racal's Radio Group. Other partners included Millicom, which had operated cellular phone systems in the USA – including a test-rig around Rayleigh-Durham, North Carolina, as early as 1981. Racal's group decided to trade under the name 'Vodafone'.

A committee, the Joint Radiotelephone Interfaces Group, on which all the governmental and business parties were represented, decided which cellular standard to adopt. The Nordic NMT was rejected because it would not provide enough capacity for central London. Other possibilities were ruled out for being either proprietary or unproven. In common with much policy during the Thatcherite 1980s, eyes turned to the USA for inspiration. The American AMPS was a proven standard and would have been ideal, but it operated at frequencies already occupied in the UK. So a tweaked standard, based on AMPS, rather grandly called Total Access Communications System, or TACS, was quickly agreed.

The first cellular phone call in the UK over the new services was made on New Year's Day, 1985, fittingly from St Katherine's Dock, in the City of London, to Vodafone's headquarters in Newbury, fifty miles west in the Berkshire countryside. (One of the callers was the comedian Ernie Wise.) Cellnet launched in the same month. However, neither Cellnet nor Vodafone was an instant success. In a final desperate, and successful, appeal to market forces, a further layer of competition was introduced. Service providers, small entrepreneurial firms, took over the task of selling cellular phones to the public. With the instincts of a barrow-boy trader, these easy-come, easy-go firms aggressively pushed mobile phones to punters. Some private fortunes were made. Many providers would later be swallowed up into more respectable groupings, such as Carphone Warehouse and Hutchison Telecommunications. But by then the cellular phone provided a growing business for Vodafone and Cellnet, and symbolised the 80s ethos of competition. The brick-like cell phone clasped to the ear became part of the conspicuous consumption exhibited by the City high-flyer – in cliché, the trappings of the yuppy.

The marketing of mobile phones in the 1980s

A young urban professional uses a Telecom Coral cellular phone as a driverless train arrives on the Docklands Light Railway. The chunky mobile phone and the development of the Canary Wharf site in East London were entwined symbols of the Thatcher years. (By courtesy of BT Archives.)

was aimed squarely at business people. The following exhortation from 1986, from the instructions issued to the salesforce for British Telecom Mobile Cell phones, is especially revealing.

TURNING IDLE TIME INTO PRODUCTIVE TIME

When you're away from your office and your phone, you're effectively out of touch with your business. You can't be contacted. Nor can you easily make contact yourself. Take a mobile telephone – a Cell phone – with you and you get a double benefit. You're totally in touch, ready to take instant advantage of business opportunities when and where they occur. And you can make maximum effective use of 'dead time' – time spent travelling – turning it into genuine productive hours.

A society in constant touch was partly created by this economic rationale to squeeze ever higher levels of productive work. Resurrecting 'dead time' – the phrase could equally well refer to time spent with families or at leisure – and reclaiming it in the service of capital.

In 1989, competition for three more licences was announced. While it was hoped that these services, called Personal Communications Networks (PCNs), would be highly distinctive – more down-market

The end of 'dead time'. A businessman is still at work while taking a London black cab, 1985. (By courtesy of BT Archives.)

than TACS, and cheaper than the GSM underway across the English Channel – in fact, to all intents and purposes, they were indistinguishable. (The PCN proposals were loathed by mainland Europeans. It was seen by the Germans and French, in particular, as an affront to the spirit of European co-operation represented by GSM. The eurosceptic Margaret Thatcher was happy with this divergence.) Only eight PCN bids were received. It was by sheer chance that the bidding for the untried PCN and the potentially lucrative German GSM licences coincided, and many international firms understandably concentrated on the surer bet.

One licence was already promised to Mercury (in fact, a grouping of Cable & Wireless, Motorola and the Spanish telecoms monopoly Telefónica), so that it could continue in its efforts to compete with British Telecom. Mercury's service was called 'One 2 One'. The other two winners were consortia called Microtel and Unitel. The losing consortia included some of the big names of British electronics, such as GEC, Ferranti (in the form of a spin-off, Ferranti Creditphone) and Kingston Communications, the innovative private telecoms company based in Hull. In 1991, a bewildering sequence of changes in consortia ownership led to just two groupings: a merged Mercury and Unitel, owned by Cable & Wireless and US West, and offering One 2 One; and Microtel, owned by Hutchison and the defence firm British Aerospace (BAe), soon to rename itself 'Orange'.

Mercury One 2 One launched in London in September 1993. So when Orange followed in April 1994, four different cellular phone services were offered to customers in the UK. While the two older analogue TACS (Vodafone and Cellnet) and the two newer digital PCN systems (Orange and One 2 One) were technically distinct, the service was very similar and the customer made little technical distinction between the four. (Indeed, all were digital after the mid-1990s.) The result was intense competition based on billing packages and sharp advertising. A revolutionary shift came with the offering of 'Pay As U Go' packages, which were simple, required no credit check and were anonymous. One 2 One offered off-the-shelf packages with pre-charged batteries – ideal as a gift or for those daunted by complex tariffs. Vodafone promised 'Pay as you talk'. Many of the deals were stoked by the marketing tactic – reprehensible in the view of many mainland Europeans – of selling telephones at very low prices, subsidised by air-time revenues.

But marketing was not pursued by price alone. While all the network operators invested heavily in advertising, it was Orange that made the early running, building a trusted brand, with a catchy slogan – 'The future's bright, the future's Orange' – and breezy amiable ads. Orange also boasted per-second billing and – most important for the fashion-conscious urbanite of the early 1990s – sleeker Nokia phones. The other operators took time – and money – to catch up. Star celebrities helped to sell phones. For example, in 1996, the supermodel Kate Moss was wishing for a 'one to one' with the young Sun Sessions-era Elvis Presley. (Orange had commissioned extensive public opinion research from the MORI organisation, and had found out that Elvis was 'one of the most popular famous people'.)

Meanwhile, all four networks piled on new clients. By the summer of 1994, One 2 One had connected its 100,000th customer. The millionth signed up in January 1998, the 5 millionth in April 2000, and the 9 millionth by 2001. Cellnet passed 100,000 in 1988, 1 million in 1994 and 11 million by 2002 (by which time the old BT spin-off had puffed itself up with the groovier name 'O_2').

Once the expensive business of rolling out the cellular network infrastructures – the base stations, switches, microwave links and so on – had been completed, the revenue from customers rolled in. The cell phone companies soon became industrial giants. In August 1999, One 2 One, less than a decade old, was bought by Deutsche Telekom, becoming part of the worldwide T-Mobile roster of networks, for £8.4 billion. ('T-Mobile' was a contraction of DeTeMobile, the name that had been given to Deutsche Telekom's mobile activities in 1993.)

Anglo–German commercial interchange continued the following year when Vodafone, less than two decades old and already combined with the American AirTouch Communications, acquired the German technology and media giant Mannesmann, making it one of the largest companies in

Europe and one of the top ten companies, by market capitalisation, in the world. As part of the same deal, Mannesmann sold Orange (which it had snapped up in 1999) to France Telecom. By 2001 Vodafone had become an international player, with more than 80 million customers across the world. Furthermore, Vodafone symbolised another significant shift. For anyone following British industry for much of the twentieth century, the idea of a British firm buying a major German technology-based company would have seemed laughable. But Vodafone's share price, riding a wave of tech-stock enthusiasm, gave it immense purchasing power. However, as we shall see, the market can go down as well as up.

Competition between One 2 One, Orange, Cellnet and Vodafone had brought prices down, and made the mobile phone an everyday object. No longer was it a status symbol – signifying privilege in the 1950s or wealth in the 1980s – but instead the universal accompaniment of young and old alike (although particularly the young). As the mobile slid down the social scale, it became a great leveller: granting the power of mobile communication and organisation to the shifting, roaming crowds. A population in constant touch. Indeed, there was an ironic twist in the levelling powers of the mobile phone in 1992. In August, the tabloid newspaper

The Sun devoted ten pages to a taped conversation between Diana, Princess of Wales and James Gilbey, revealing the two to be lovers. The royal marriage, already strained, was in tatters. In December, in the same month that Windsor Castle burned, the prime minister, John Major, announced the divorce of the Prince and Princess of Wales to the House of Commons. The source of *The Sun*'s revelations, the so-called 'Squidgy' tapes (named after Gilbey's pet name for Diana), had been recorded by eavesdropping the Princess's mobile phone.

· CHAPTER 9 ·

DECOMMUNISATION = CAPITALIST POWER + CELLULARISATION

In 1920, Vladimir Ilyich Lenin, leader of revolutionary Russia, surveyed a country ruined by civil war and racked by starvation, and confidently announced that progress was assured through the rapid construction of a technological network. His slogan was pithy: 'Communism is Soviet power plus the electrification of the whole country.'

With the fall of communism across central and eastern Europe seven decades later, the nascent liberal democracies were blessed with a small army of economic advisers from the West. Their mantra, heard loudest in Lenin's homeland, called for the unleashing of entrepreneurial activities, the rapid privatisation of state-owned industries and the opening of markets to foreign companies. Following close behind the economists were western cellular phone companies.

We have already seen how mobile phone systems were built in the former East Germany as a way of

Did you join the network? (With apologies to Dmitry Moor.)

providing communication services without relying on obsolescent land lines. However, the rise of capitalist power in Russia did not coincide with the cellularisation of the *whole* country. Not only was it too vast, but also socialist ideals such as universal coverage had been abandoned. A condition of gaining a licence, such as those enabling US West and Millicom's services in Leningrad and Moscow (both from 1991), was that local partners were involved. In the capital, licences were granted, sometimes for cash donations to the local powers that be, for a number of different standards.

Yet again, the style of the technological system was being shaped by political context – in this case, the turmoil and confusion of post-Soviet Moscow government. At one stage Bell Canada thought it had won a GSM licence, but then backed out when a surprise demand for US$50 million was made. Likewise, as Garrard records, it was to the 'constern-ation of operators that had received NMT450 or GSM licences [that] AMPS licences were announced for Moscow and three cities in the East of the country'. One of the Moscow licences went to a consortium led by the Cold War defence giant Vimpel. In 1996, VimpelCom, founded by Vimpel's Dmitry B. Zimin and American investor Augie K. Fabela II, became the first Russian company listed on the New York stock market. 'Bee-line', Vimpel-Com's mobile brand, symbolises a new Russia, in which western investment has put flesh on old Cold War bones.

Likewise in China, the transition from

communism to some form of capitalism was being reflected by the spread of mobile base stations. Chinese cellular telephony started in 1988 with a TACS system (that is, the British standard) in Beijing, Shanghai and Guangdong province. Two companies, China Mobile and China Unicom, offered increasingly popular GSM-based services in the 1990s. By 2002, China had become the number one mobile country: there were 160 million subscribers, overtaking the USA. Measured by percentage of population, of course, the picture looked somewhat different (just over 12 per cent of Chinese compared to almost 38 per cent of Americans had cell phones). The potential size of the Chinese market was clearly huge.

However, western companies attracted by the market often met a rocky reception. It took nearly a decade of lobbying, for example, for Qualcomm to gain permission to supply a CDMA network to China Unicom, negotiating past tricky moments in China–US relations, such as the bombing of the Chinese embassy in Belgrade in 1999 and the ramming of a US spy plane by a Chinese jet in 2001. But Qualcomm's determination seems to have paid off, and the company hopes to follow Ericsson's example: the Swedish firm's biggest market is in China, not Europe or the USA.

· CHAPTER 10 ·

JAPANESE GARDEN

It might seem strange to have read over half a book about the development of a new consumer technology without Japan entering the story. In fact, for the first decade of the Japanese cellular phone, there was little out of the ordinary. Indeed, it paralleled developments in an average European country, Spain say, where a monopolistic public telephone service introduced an early but expensive system, and take-up rates were low. Nippon Telegraph and Telephone (NTT), then a public corporation, launched a cellular service around Tokyo, and a smaller one around Osaka, in 1979 – among the first commercial cell phones in the world. But ten years later, only 0.15 per cent of the population had bought the deal.

Of course, the peerless Japanese electronics manufacturers, such as NEC, Matsushita (under the Panasonic brand name) and Sony, supplied equipment to cellular services growing elsewhere in the world. But despite the excellence of the product, export suffered from measures taken to protect American or European markets. By the late 1980s, many telecommunications markets were being opened up to competition. However, regulation could take protectionist forms, even if it had been designed for other reasons. In the case of GSM phones, the thorny and complex patent problem meant that just a handful of companies (of which only Motorola was non-European) gained access to a highly profitable pan-European market. Japanese firms lay outside the walls of the 'fortress Europe' for phones. They were also hampered by the bursting of the Japanese bubble economy in the early 1990s, which meant that the manufacturers were distracted, inward-looking and cautious, just as GSM was launched in Europe.

The Japanese mobile gets interesting – and very distinctive – from the late 1980s. NTT's monopoly was broken when two companies offered cellular services: the Nippon Idou Tsushin (IDO) corporation around Tokyo, and Daini Denden Incorporated (DDI) around Kansai. The resulting patchwork of incompatible standards bears comparison with Europe before GSM or the USA even now. While GSM represented for Europe a remarkable experiment in co-ordinated technological action, this style of policy, involving close co-operation between government agencies and the private corporations, was for Japan a familiar and successful model of technological innovation. The result was a new digital mobile standard, agreed by 1989, and licensed to three consortia, led by NTT, Nissan and Japan Telecom respectively. With compatibility and competition came fast growth – nearly 9 million subscribers (7 per cent of the population) by the end of 1995.

It was NTT's mobile division, later a separate company eventually called NTT DoCoMo, that benefited from an astounding transformation of the mobile phone, a change led by the users themselves. The company had already built up a large customer base for its cellular phones: 1 million subscribers in 1993 (the year of the digital launch), 10 million by February 1997 and 20 million only eighteen months later. (And the growth would continue: 30 million in 2000 and 40 million by 2002.) But in 1999, DoCoMo created a new service. Called 'i-mode', it was an instant hit. Distinctively, it was the hyper-fashion-conscious Japanese teenagers that led the way. Ten million had i-mode by 1999, a number that doubled in a year and trebled in two. They wanted i-mode, but what was it?

When I smashed my phone up, I was amazed how many components were inside. There's a lot more than just a microphone, loudspeaker, radio

The DoCoMo 'Pacty', 2001. This device illustrates
several distinctive features of Japanese mobile culture,
in which image and design are highly valued: it is
styled to appeal to the young female consumer, and
greater attention is given to good graphical content.

transmitter and receiver. Most important, there's a
chip. The presence of a microprocessor means that a
cell phone can do anything that a computer can do:
it can send, receive, store, show and change data in
pretty much any form required. So there was never
any reason to think of it as just a voice communication device. A mobile is – or could be – much,
much more than a phone.

So a mobile could carry data as well as mere voice.
In the 1990s, this was already old hat. But while

speculators might pin their hopes – and invest-
ments – on a convergence of information and
communication technologies (which produced the
great telecoms bubble that would nearly wreck the
global economy in the first years of the twenty-first
century), there was precious little evidence from
anywhere in the world that consumers wanted
more than a phone. Videophones, for example, in
which you could see who you were talking to, have
been technically feasible for decades and launched
several times, but had always failed. Dashing the
hopes of telephone companies, people preferred the
low-tech, low-bandwidth voice-only telephone
conversation, with its capacity to keep revealing
expressions – and one's appearance – hidden.

Text-messaging had been written into the GSM
standard, but almost as an afterthought. In the
West, the attempt to combine limited access to on-
line information via the mobile phone, based on
Wireless Application Protocol (WAP), was greeted
with much fanfare in early 2000, but was a
disappointment. The limited range of WAP content
available made the service seem tame compared
to the wilds of the Internet, and dispiritingly
corporate. For those who wanted the corporate
product of Disney, Nike or Champions League
Football – and there were, of course, millions – there
were better ways of consuming it than via the inch-

wide WAP screen. For those for whom the restriction to corporate product was a turn-off? Well, they never turned WAP on.

i-mode was like WAP, but it worked. In both, the operator controlled access to digital content. But while the walled garden of WAP wilted, i-mode's flourished. To understand why, we need to look closer at the horticulture: how i-mode content was grown and provided, and how it fitted with Japanese society.

i-mode acted as a gateway. Any provider of content had to satisfy NTT DoCoMo's strict rules concerning what content was allowed and what was not. Content would be current, attractive and safe. This guarantee of quality meant that the owner of the i-mode phone would not be shocked or defrauded. In return for self-censorship, the content provider was granted access to NTT DoCoMo's already considerable market of subscribers. Furthermore, the complicated question of billing – the rock on which nearly all models of electronic commerce are wrecked – was solved: amounts of downloaded data were totted up and charged as part of the phone bill. Only DoCoMo, gatekeeping transactions as broker and guarantor, could do this. Furthermore, unlike WAP or most people's personal computers (for the Internet), i-mode would be always on-line. Kei-ichi Enoki, managing director for i-mode, sums up i-mode's philosophy of restricted content and constant access:

> *PCs are like department stores. They have a wide selection of content, including excellent graphic images. If you decide to make a visit, you can stay as long as you like and explore different sites at your leisure.* [By contrast] *mobile phones are more like convenience stores, where only a selection of goods are on display in the limited space available. The contents have to be simple, but the convenience comes from the fact that they can be accessed at any time.*

Convenience was, of course, a virtue in an intensely hectic country such as Japan. Indeed, when i-mode was launched, the content sanctioned by NTT DoCoMo was squarely aimed at the busy salary-man. For ¥300 a month, a January 1999 press release promised subscribers the chance to reserve airline and concert tickets, check their bank balances, transfer money, send and receive e-mail (a real draw, since government regulations of NTT had checked the growth of the Internet in Japan), and to register for the 'Message Service', which automatically provided information on 'weather, stocks and other topics, depending on choice'. All very safe, adult and probably heading for failure.

Instead, i-mode was discovered by teenagers and young adults. In particular, i-mode appealed to single women in their 20s, whose economic status was rising rapidly, just as that of the salaryman with a job for life declined. A different, and very profitable, dynamic sprang up. Not only did i-mode give its young subscribers instant access to new goods, but also it kept them in constant touch with each other, intensifying the pattern of rapid stylistic change still further. In turn, providers of commodities, well aware that the spending power of the Japanese teenager was second to none, had access through i-mode to a fast-moving, capricious, but profitable, market. The young yen saved i-mode, and demonstrated that the mobile phone could really be more than just a talking point.

PART III

MOBILE CULTURES

· CHAPTER 11 ·

TXT MSGS + TXTPOWER

Txt Msgs

Wot is kltr? Kltr is the clln of sgns spcfc to a sosIET. evry tek hs a kltr of its own. a kltr cn b hrd 2 undrstnd 2 outsdrs. ther is no bttr illstrn of ths thn txt mesgs.

Txt msg ws an acidnt. no1 expcted it. Whn the 1st txt mesg ws sent, in 1993 by Nokia eng stdnt Riku Pihkonen, the telcom cpnies thought it ws nt important. SMS – Short Message Service – ws nt considrd a majr pt of GSM. Like mny teks, the *pwr* of txt – indeed, the *pwr* of the fon – wz discvrd by users. In the case of txt mssng, the usrs were the yng or poor in the W and E.*

TxtPower

City life is mobile. City life is fast. And in no region – outside the Nordic countries – have cellular phones become as culturally important as in the cities of the Pacific Rim. In entrepreneurial Hong Kong, where, writes Garrard, 'it is almost as important to look busy and important as it is to make a deal', there has never been enough capacity to meet demand for mobile phones, despite licences being granted for any and every standard. Hong Kong needed six DCS 1800 services, for example, when the UK had been satisfied with just two (Orange and One 2 One).

When China took back Hong Kong from Britain in 1997, the mobile genie had escaped the bottle and the strict controls over telecommunications found on mainland China would have been nearly impossible to introduce. Likewise, Australian cellular phone systems were centred on flourishing cities such as Sydney and Melbourne. Indeed, so strong is the expectation of being able to keep in constant touch via the mobile phone that inexperienced travellers into the Outback have to be repeatedly warned that the cellular signal dies a few miles outside of the major towns. But while the cell phone has bolstered the existing entrepreneurial culture of Hong Kong and confirmed the sociability of Australian city life, in the Philippines the mobile led to a political revolution.

Under Joseph Estrada, who had been elected president in 1998, many Filipinos felt that the country was slipping back to the corruption and cronyism of the Marcos days. But whereas the old dictator could introduce martial law and crack down on opposition, and the anti-Marcos coalition had to rely on ham radio and mimeographed pamphlets, the opponents of Estrada had a new tool in their possession. Mobile phones in the Philippines took a while to take off, mostly because spectrum space had to be cleared – a complex task when unofficial radio was a major means of contact in an island archipelago with patchy or absent telecommunications infrastructure. But in 1995, Smart Communications, a consortium that included First Pacific and NTT, entered the market, bringing much cheaper cellular service prices. By 1996, 7 million (10 per cent of the population) owned a mobile phone – almost twice the number with land lines. Perhaps more importantly, pre-paid cell phone services were readily within the reach of the poor.

Across the world a split had developed by the late 1990s between those who paid for mobile phones by monthly bills (and who therefore registered personal details and submitted to credit checks) and those – the young, the poor – who used pre-paid services, usually by purchasing top-up cards, with a lower starting cost (but often higher call charges), and who, as a result, remained anonymous. Text messaging was encouraged by the use of top-up cards, since to eke out the minutes it was better to use fractions of a second sending a text than waste whole minutes in conversation. (Indeed, texting was sometimes made free as an enticement to new customers.)

In the USA, text messaging was not popular, since phones were incompatible and the cost advantages mattered less to the affluent. (Additionally, beepers and pagers had a prevalence unmatched elsewhere in the world.) As a result, mobile culture is far less rich in the USA than in texting hotspots such as Finland, Italy, the UK or, particularly, the 'text capital of the world', the Philippines. According to Rodolfo Salalima, the vice-president of leading Filipino carrier Globe Telecom, about 80 per cent of his company's customers used pre-paid cards. The cheapest phone card cost about US$5 and was good for two months. In an agricultural, Catholic country where the extended family was important, but also in an industrialising country where the young were drawn or forced to the cities, US$5 bought cohesion. The outcome was that by 2001, not only did the Filipino élite communicate by cell phone, but also the rest of the population, vast and previously poorly connected, possessed anonymous text-message-enabled phones.

Text messaging played a key role in ousting Estrada. In late 2000, rumours spread as quickly as fingers could text true, exaggerated and imagined stories of Estrada corruption. Over 100 million text messages flew around the Philippines each day. It started with jokes such as: 'The NPA [Communist rebels] have kidnapped Erap [Estrada's nickname; it means 'buddy' backwards in Tagalog, the Philippines' main language]. They are demanding a large ransom and, if it is not paid, they are threatening to release their hostage.' (To illustrate the indiscriminate power of text, another hoax announcing the death of the Pope was also passed on by millions.)

As push came to shove, people were texted: 'edsa. edsa: everybody converge on edsa' – Edsa being the shrine that was the focus of the challenge to Estrada. While it was only after his cabinet had defected to the opposition, and the army and the police had transferred allegiance, that Gloria Macapagal-Arroyo was swept to the presidency in January 2001, it was also 'People Power' brought together by text messaging that forced them to shift.

Macapagal-Arroyo immediately acted to ban 'malicious, profane, and obscene' texts, which offered some protection against her predecessor's fate, but she has not been allowed to forget the power of text. In September 2001, Smart Communications and Globe Telecom announced that free texting would be reduced. Immediately, Txtpower, a group formed by cellular phone subscribers, organised the sending of at least 1 million text messages to President Macapagal-Arroyo to urge her to intervene to save free text messaging and to act on the alleged 'lousy' services of the phone firms. Likewise, when in May 2002, President Arroyo announced plans for the taxing of text messaging – the country's public debt was US$50 billion, or 70 per cent of gross domestic product (GDP) – Txtpower and sympathetic politicians reacted angrily. For them, free texting equated to freedom of Filipino expression.

The story of the Philippines shows, once again, that mobile phones are moulded by the countries they are used in and give form to the nation in return, but it also acts as a reminder of another theme of this book: the shift away from centralised hierarchical modes of organisation towards decentralised networks. This was not driven by technological change, although new technologies, of which the mobile phone and the Internet were prime examples, symbolised and supported it. Instead, they reflected the great shift in models of governance – that is to say, the ways in which decisions were taken and acted on by organisations, be they government, industry or non-governmental organisations.

CHAPTER 12

TWO ORGANISATIONS IN THE CONGO

In early 2002, Mount Nyiragongo, one of Africa's most active volcanoes, erupted. This time, however, the lava flow passed through the centre of Goma, a Congolese city on the edge of Lake Kivu near the border with Rwanda. In the 1990s, it had been the temporary home of the Rwandan Hutu refugees, many responsible for genocide, as well as the centre for the Rwandan-backed rebellion against Laurent Kabila's Congolese regime. Many who were part of the second wave of refugees, leaving Goma to escape from the advancing lava, recalled the miserable experience of the Hutus and were determined to stay in the camps for as short a time as possible. Around Goma, therefore, two styles of organisation mattered.

The first was represented, in the absence of the distant Kabila government, by the United Nations High Commission for Refugees and the charities. These were organisations with clear centres, distributing aid along chains of command. The second was formed from the centre-less networks of gossip, given technological form by the mobile phone. As the news channel CNN reported: 'An unlikely adversary has emerged in the battle to bring relief to the victims of the Congo volcano tragedy – the mobile phone.' Oxfam worker Rob Wilkinson said that while aid agencies were telling people not to return to the city of Goma and to stay in the refugee camps, mobile phone calls were persuading them to return. 'They are using mobile phones to talk to relatives and friends back in Goma, who are telling them that it is OK to go back', he told the Press Association. 'It is changing the way the population is responding. It's very unusual.'

While confusing, what Wilkinson had witnessed was an increasingly common phenomenon in many areas of life: a weakening of centralised hierarchies in the face of strengthening networks.

· CHAPTER 13 ·

THE NOKIA WAY – TO THE FINLAND BASE STATION!

The case of Goma and the volcano reminds us that a key aspect of mobile culture is, perhaps obviously, mobility. (And a mobility that cuts across national boundaries.) But there is also a distinctive *material* aspect to mobile culture, which is best illustrated by the products of the phenomenal Finnish company, Nokia. While material culture might seem at first to be mere flotsam and jetsam, and not part of the great tides of history, I think the opposite is often the case. Indeed, something as trivial as a coloured plastic phone cover – called a facia, or fascia – can be as much a vehicle of grand historical change as fascism.

Without doubt, Nokia has become the most influential manufacturer of mobile phones in the world. But why did such a firm emerge from Finland? Industrialisation came late to the country on the northern fringe of mainland Europe, but when it did so it made use of one natural resource that Finland, like Sweden, had in abundance: forest. In 1863, after a daring act of industrial espionage, Knut Fredrik Idestam imported a new wood-pulp process from Germany and set up a mill on the Nokia river, a few miles outside the small city of Tampere.

Right from the start, the enterprise had close links with Finnish politicians. Idestam's partner was Leo Mechelin, a parliamentarian and financier who helped to extricate Finland from its status as a Russian duchy to being an independent state. For much of the twentieth century, Nokia was an industrial coalition between pulp, rope, cable and rubber works. Indeed, as Nokia's historian Dan Steinbock records, as late as the 1980s Nokia simultaneously brought electricity to 350 Egyptian villages, made most of the toilet paper in Ireland and provided all the studded bicycle tyres in the world.

But the peculiar position of Finland in world politics meant that Nokia was quite unlike any other European conglomerate. First, ever since the Russian revolution, Finland had had to play a delicate balancing act between the capitalist West and communist East. This strategy, a combination of cautious neutrality and *realpolitik*, is known as the 'Paasikivi-Kekkonen' line (after the two politicians who adopted it). For example, Nokia's major market was the Soviet Union (not least supplying many of the power cables for Lenin's and Stalin's programmes of electrification). But later, with some prescience, Nokia's boss, Kari H. Kairamo, decided that such dependency should be balanced by building up West European links. The policy had been echoed at national level when Finland, after delicate negotiations between East and West, joined the European Free Trade Area (EFTA) in 1961 and signed trade agreements with the EEC.

Second, Finland was extremely dependent on imported oil, which had to be paid for by increased exports, which again gave cause for good trading relations, East and West. So, under Kairamo from 1977, Nokia sought the means to create innovative exportable electronic products.

Finally, when the telephone came to Finland, it was not – unlike any other country in Europe – placed under the control of a single monopolistic operator, but under a host of independent local operators instead. There were over 800 in 1938, and still around fifty in the 1990s. Again, the cause can be found in Finnish foreign relations. The distinctive Finnish telecoms pattern of links between private companies and many local co-operatives, was, notes Steinbock, due not so much to 'boosting the efforts of the private sector as trying to keep Russian authorities away from the emerging industry' (a strategically important industry at that). The important consequence for Nokia was that it had on its doorstep a diverse market for competitive products, and that it never had to compete with a big national telecoms monopoly.

Nokia already manufactured a few mobile phones at its Oulu plant in the far frozen north of the country, when Kairamo signed a deal with the Finnish TV manufacturer, Salora Oy. The joint venture, Mobira Oy, begun in 1979, was soon owned outright by Nokia when Salora was swallowed up in 1984. Also in this period, and for reasons that are unclear, Kairamo tore down Nokia's hierarchical organisation, typical of many a European conglomerate, and replaced it with a decentralised 'flat-pyramid' management.

This radical change, which would only later become a new orthodoxy of managerial science, seems to have been based on Kairamo's shrewd analysis – or guess – that the world system of two superpowers was nearing its end and only a nimble company would be able to exploit the new global opportunities. What is certain is that, if the tremors of the coming earthquake *could* be felt, then Finland, balanced precariously between East and West, was near the epicentre.

Kairamo, a manic depressive, hanged himself in December 1988, having become convinced that a forced break-up of Nokia was imminent. In fact, the restructuring that was initiated by Simo Vuorilehto and completed by Jorma Jaako Ollila, was built on Kairamo's legacy. Ollila was the person most responsible for focusing Nokia almost entirely on mobile phones. In effect Mobira Oy, along with a few other electronics and cable divisions, became the whole company. Many factors had combined around 1990 to permit this. Finland had a purely conservative government, intent on telecoms deregulation, for the first time in decades. The Finnish economy was in a tailspin following the collapse of Soviet trade, and a new direction was needed. This direction clearly pointed towards further European integration (Finland joined the European Community in 1995), of which GSM was to be the showcase of pan-European potential. Indeed, the pro-European emphasis had already prompted Finnish involvement with the Nordic NMT standard and in early GSM discussions. So Ollila bet the company on mobile phones.

But if placing Nokia in a political context helps us to understand why it was in a position to become a major mobile phone manufacturer, we need to go a bit further to account for its extraordinary success based on distinctive products – the material culture. Nokia's mobile phones of the 1980s such as the Cityman (1986), which was very popular in the UK,

Jorma Jaako Ollila, Chairman of the Board and CEO, Nokia. (Nokia Photo Archive.)

already boasted superior design. Styling and brand were more important to Nokia than they were to competitors, such as Motorola or Siemens, say, or at least were pursued with greater success. (Again, there was a cultural advantage: good industrial design was the material analogue of Nordic social welfarism, the sharing of the benefits of industrial society through rational planning.) But let's take just one object – to my mind, an iconic one – to see what it reveals about the Nokia way.

The Nokia 3210 was launched in summer 1999. It is a design classic. 'Elegantly styled, with no protruding aerial and lovely slim proportions', drooled

The Nokia 3210 was to cellular communications what the Ford Model T was to the automobile. (Nokia Photo Archive.)

What Cellphone. I, too, recall the thrill: the silver and grey phone seemed moulded to fit perfectly in the hand. It was obviously immediately an icon, like Coca-Cola's bottles or the Citroën DS. (And by 2020 the sight of a Nokia 3210 will trigger millennial nostalgia.)

In one sense, the 3210 did for mobile phones what the Model T Ford did for the automobile: it was a cheap, but beautifully engineered, vehicle for mass communication. But if the Model T Ford symbolised the dominant style of production of much of the twentieth century, the Nokia 3210 represented its opposite. Fordism stood for centralised control, hierarchical management and, famously of the Model T, 'any colour, as long as its black'. Nokia boasted flexibility of production, flat hierarchies, and products that reflected this organisational style. With the Nokia 3210, you could change colour simply by choosing a new 'Xpress-On' facia. *What Cellphone* relayed the reaction of Janice Caprice, a London beauty therapist:

> *It's got to be eye-catching, anything from the British flag to a flower. Most of my friends buy a phone because they can get a cover for it. I bought an Ericsson PH337 for that very reason but that's old now so I'm saving up for a Nokia next.*

Nokia had experimented with Xpress-On with the slightly earlier introduction of its more conventional 5110 phone to market in 1998, and other manufacturers had aped the innovation. But with the 3210, interchangeable facias became integral to the product's design and marketing. Facias are superficial and shallow. But they are also colourful and flexible, and mean that the same phone can display different allegiances, as fashions shift. I think we should take such superficiality seriously. The 3210, like the 5110, carried simple games derived from earlier classics (Snake, Memory and Rotation), more evidence of the incipient shift from mobiles as mere communicating devices to something more. The 3210 was also the first Nokia phone to carry T9, a predictive text system developed by a small company called Tegic, which shoehorned the equivalent of a full alphabetic keyboard on to just a number pad. The phone contained a dictionary and software for searching it, which was possible because the mobile contained a microprocessor.

The contrast between the squat black Type 300 Post Office phone from my childhood and the chameleon-like 3210 should be clear. To hold the 3210 in the palm of your hand is to have evidence, in material form, of a great transformation.

· CHAPTER 14 ·

MOBILE PHONES AS A
THREAT TO HEALTH

Stories of bodily harm caused by mobile phones were commonplace across the industrialised world by the late 1990s. The stories were directed against two culprits – radiation from base stations and radiation from the mobile phones themselves – but the tones were very similar. A story that could be found in the *Daily Mail*, a British middlebrow, mass-circulation newspaper, in December 1999, was typical. Under the headline 'Now mobiles give you kidney damage', the reader was told: 'Scientists say exposure to the phones' low-level radiation causes red blood cells to leak haemoglobin. The build-up of haemoglobin, which carries oxygen around the body, can lead to heart disease and kidney stones.' The reader would already have known of earlier stories suggesting links between mobile phone use and brain cancer, premature ageing, diseases such as Alzheimer's and Parkinson's, multiple sclerosis and chronic headaches.

A base station aerial being erected in 1985. Two
components of cellular phone systems provoked
anxieties over adverse health effects: the handset and
the base station. Base stations near, or on top of,
schools caused the greatest concern. (By courtesy of
BT Archives.)

Chilling stuff. And although there was the usual disparity between headlines ('Now mobiles give you kidney damage') and research ('More work is needed to investigate some results which seem to indicate that electromagnetic waves in the radio spectrum may interfere with processes within the kidney'), the economic importance of the mobile industry forced governmental organisations to act. In the USA, regulation of cell phones is shared by the Federal Communications Commission (FCC), which sets guidelines concerning levels of radio frequency (RF) radiation, and the Federal Food and Drug Administration (FDA), which has a brief to follow health matters. The FDA set out to reassure cell phone users that the technology was safe.

In Britain, the Department of Health might have played a similar role to the FDA. However, with fiascos such as BSE in the recent past, the government chose to ask an independent expert group on mobile phones to investigate. The group reviewed media coverage, and from September 1999 heard evidence from scientists, members of the public and representatives of the telecoms industry and special interest groups (for example, Powerwatch, Friends of the Earth, Scotland and the Northern Ireland Families Against Telecommunications Transmitter Towers).

The findings, called the Stewart report after

the group's chairperson, the biologist Sir William Stewart, were published in 2000. The balance of evidence suggested that exposure to radio frequency radiation at levels below existing guidelines 'did not cause adverse health effects'. However, the Stewart report went on to say that 'there may be biological effects' at such levels, and therefore it was 'not possible at present to say that [such] exposure ... is totally without potential adverse effects, and that gaps in knowledge are sufficient to justify a precautionary approach'. In particular, children should not be encouraged to use mobile phones because their bodies were still developing.

The Stewart report's conclusions were more cautious than other governments' investigations. The Health Council of the Netherlands, for example, concluded that there was 'no reason to recommend that mobile telephone use by children should be limited as far as possible'. But such reports were also shy of making strong general claims over the safety of mobile phones. The World Health Organization (WHO), on one hand, records that 'none of the recent reviews have concluded that exposure to the RF fields from mobile phones or their base stations causes any adverse health consequence'. Yet on the other hand, it felt the need to rush out press statements correcting press articles which reported that the WHO had insisted 'mobile phone emissions are safe'. The billion dollar insurance claims over damage caused by asbestos and tobacco have made all organisations wary of putting their name to pronouncements of complete safety.

There have been many attempts to close the debate over the health effects of mobile phones. But, as in other controversies such as BSE, or Mad Cow Disease, experts tend to disagree rather than agree and closure is rarely attained by expert pronouncement. The issue is also necessarily open-ended, since it is impossible to say what length of time is sufficient to be satisfied that long-term harmful effects do not exist. Nor will a technical fix soothe fears. A small industry has grown up offering technical solutions, from headsets (so the phone irradiates your guts rather than your head), to fraudulent quack remedies involving 'absorbent' phone covers. These products exist *because of* anxieties, not to allay them.

Rather than expect the debate over health and mobile phones to be resolved, two quite different ways of thinking about the debate should be considered. First, with subscription to mobile phones hitting over three-quarters of the population in many countries, the big picture is not of users resisting a technology, but of enthusiastically embracing it, despite knowledge of risk. What needs to be explained is not so much why there is public concern over harmful effects of mobile phones, but why the concern has so little effect over behaviour. Second, the debate will never be closed by expert pronouncement, since public concerns are framed by a powerful and growing trend of distrust towards scientific expertise. (I suggest this trend is part and parcel of wider social and political transformations, discussed later.) Concerns supposedly directed at a particular technology are, in fact, generated by deeper social tensions and conflicts. Indeed, expert opinion is united to declare harmful effects only in one area: talking on a mobile phone while driving. This opened a new attack on an old alliance between two technologies of mobility.

The history of two technologies of mobility and individual freedom, the car and the mobile phone, has long been intimate. (By courtesy of BT Archives.)

· CHAPTER 15 ·

CARS, PHONES AND CRIME

New technologies of mobility create new crimes, new criminal *modus operandi* and new ways to catch criminals. Take, for example, the automobile. From the first decade of the twentieth century, cars provoked a crime wave. In Britain, a speed limit of twenty miles an hour was in force between 1903 and 1930, when it was briefly revoked. A thirty mile an hour limit in built-up areas was hurriedly reintroduced by the 1934 Road Traffic Act after a spate of accidents, and has been in place ever since.

Mobility increased through the century. By 1950 an average day's travel was five miles, but by 2000 it was twenty-eight miles. Much of this travel was by car. A dramatic effect of the new mobility, with its legal limits, was to create new crimes of speeding. Criminal statistics became dominated by car-related crime. Furthermore, as well as driving infringements becoming criminal offences in their own right, speedy automobiles led to more accidents, including 'hit-and-run' incidents. Patterns of criminal behaviour also changed. More cars meant more stolen cars. Burglars, previously confined to large towns and cities, suddenly found rich pickings in the surrounding countryside, which was now only minutes away. If you compare a graph of break-ins, you find it closely follows a graph showing car ownership.

But, in turn, techniques of policing and crime detection also changed in response to the mobility of criminals. On a mundane level, many hours of police time were now spent in work such as enforcing traffic codes or tracing missing vehicles. This new emphasis would have been unwelcome, had it not been for the new information tools that registration provided. Within a few decades of the birth of mass motoring, every driver and every vehicle had been tagged with a unique identifying number, and immense filing systems recorded changes and movements. The registers relating to motoring soon became central tools in police work.

Of course, different countries responded to the mobility of the automobile in different ways. Norway, for example, introduced strict speed limits relatively early – a pattern shared by other countries that strongly valued social cohesion and were not swayed by lobbying from car manufacturers. The USA or China would be different cases again. But the car became central to many societies in the twentieth century, and the phenomenon of increasing interdependence of policing and criminality was a common one. It should be seen as one more example of how the increased spread of technologies of mobility created possibilities of freedom from centralised authority – whether it was from the family, social institutions or other enforcers of 'correct' behaviour – which in turn were countered, even domesticated, by strengthening bureaucratic or policing powers.

A technology that demonstrated the links between mobile radio and the automobile, between cultures of mobility and freedom, was Citizens' Band (CB) Radio. CB was not a cellular radio service. All transmissions were made on a segment of the radio spectrum, around 27 MHz, that had been given over to amateur use. Everyone could hear everyone else.

From the late 1950s, a movement of enthusiasts was fostered by the marketing of CB radio kits. But CB radio really took off in the late 1960s and early 1970s, when its adoption was spearheaded by truckers, providing communication and social ties across the freeways. The CB 'fad' marked the moment when the values – and technology – of the truckers struck a deeper chord. For a few years in the mid-1970s, the specialised CB language of American truck-drivers – '10-4, good buddy',

'Breaker 1-9, this is the rubber duck' and so on –
could be heard echoed as far from Route 66 as
Britain and the Netherlands. The 1977 film *Smokey
and the Bandit* was an international smash. What
resonated was a mythology of a network of individuals living outside traditional society, indeed
outside the law, best encapsulated in the supposed
trucker's trick of using CB radio to warn each other
of police speed traps. The network opposed centralised power. While the popularity of CB prefigured
the later demand for cellular telephony, the
continuity of social values associated with both also
tells us something important.

The cellular phone is also a technology of mobile
communication, and a similar story can be told.
As with the spread of the automobile, this new
technology of mobility created new crimes. In early
twenty-first century Britain, for example, the theft
of mobile phones contributed to the largest
recorded increase in reported crimes affecting
young people. As I have already suggested, the
mobile phone is, by far, the most expensive object
ever to become routinely carried on the person.
(And by 2001, 70 per cent of adults and 81 per cent
of fifteen to twenty-four year-olds in the UK
possessed one.) Also, despite the phone's slide
down the social scale, its role as an indicator of
status and personal identity – important, as ever,

in the playground – has not been diminished. But by 2001, more than one-quarter of all robberies involved a mobile phone and nearly half the victims were under eighteen. The perpetrators too were likely to be young.

A portion of this wave of thefts by and from children stemmed from bullying rather than anything more organised. But when Home Office researchers Victoria Harrington and Pat Mayhew interviewed the inmates of Feltham Young Offenders' Institution, a more complex picture emerged. Some phones were stolen as part of the general 'pickings' of others' possessions, but expensive models might be particularly targeted. Phones would be used until the service was blocked or the pay-as-you-go card ran out. Or they would immediately be passed on to a ready local – and probably international – market. Thefts were prompted by opportunity and, intriguingly, the social tensions brought to the surface by conspicuous display of technology:

The plethora of phones around – left for instance casually on counter tops in bars, or on the shelf near the front door at home – also makes them relatively accessible to thieves. In the case of street crime, too, potential thieves can easily spot someone with a phone. It is difficult to say how much truth there is in the contention of offenders in Feltham ... that owners' ostentatious use of phones causes a degree of irritation that provokes theft – but it might give users pause for thought.

The Feltham boys expressed the same 'irritation' as the annoyed passenger on the train surrounded by mobile users, although they took rather more drastic action in response. Many police officers (although not the Home Office researchers) believed that mobile theft was often implicated in the maintenance of social status by the technique of 'taxing'. Here, 'the phone theft *per se* is less important than groups of offenders exerting control, establishing territorial rights, and showing "who's who" by penalising street users (in particular young ones) through phone theft.'

Possession of technology made social statements and, conversely, *not* to have a phone meant social deprivation. For the boys, phones were 'an indispensable crutch. Their loss of phones in custody was said to be one of the worst elements of the deprivation of their liberty.' Loss of a technology of mobility was equated directly with loss of freedom.

Reliable research on criminals' uses of mobile phones is much harder to come by, although evidence suggests that the opportunities provided by mobile phones to make planning and carrying out crimes more 'efficient' have not been ignored. A major influence has been the availability of pay-as-you-go mobile phones, which can easily be bought under an assumed name. Any calls made might be traceable, but the ownership of the phone cannot be ascertained. The same, of course, applies to stolen phones.

Before the mobile phone, the crime of cellular fraud could not have existed – another reminder that new technologies of mobility *create* new crimes. The subscription fraud form of mobile crime, in which phones are bought under false names with no intention of paying for calls, is the most frequent. But the cloning of phones, in which the identity of another user is stolen, amounted to 85,000 cases in the UK by 1996, and is widespread across the cellular world. For example, 80 per cent of drug dealers apprehended in the USA in the late 1990s were found to be using cloned mobile phones, while in the occupied territories of Israel, Palestinians made 57,000 international calls, subverting official security restrictions, and charging subscribers in Arizona!

In 2001, a complex European taxation swindle came to light, which proved even more lucrative. It was based, rather bizarrely, in Stoke-on-Trent, a conurbation in the northern midlands of England, which had not seen the good days since Josiah

Wedgwood produced pots. Fraudsters exploited the fact that no tax was charged on the traffick of goods within the European single market. They imported mobile phones, chosen because they were easily portable and in high demand, and sold them on to other, legitimate traders – adding 17.5 per cent tax, which was pocketed. At least £2.6 billion was creamed in this way in 2000–2001, with possibly £10 billion lost to the taxman overall.

Just as the mobility of the car was domesticated by keeping databases of information, the criminal potentialities of the cell phone have been contained by technologies of registration. Indeed, whereas there was no essential reason why cars and drivers had to be registered – if you don't care about the law, you can get in a car and drive off; if it has petrol it will work – the cellular phone is unimaginable without databases of information. Every time a cellular phone passes from cell to cell, a reference has to be made to some central computerised list of network users. Every time an owner is billed, the information of phone use has to be totted up and processed. Unlike the automobile, databases are integral to mobile phone technology. (However, the user might never suspect this, sometimes with unintended consequences. When itemised billing was introduced in France, male customers complained in their thousands because their affairs were uncovered overnight. The networks were forced to bow to public pressure and replace the last four numbers with asterisks.)

Most of this information is stored centrally, as part of the operation of the Mobile Switching Centre, but increasingly personal data is held in the handset. In Europe, this shift coincided with the introduction of digital cell phones. First developed for the German analogue Netz-C system, which in many other ways was a commercial disappointment, the Subscriber Identity Module, or SIM card, was adopted as part of the pan-European GSM concept. While the SIM card contained information identifying the subscriber, it also could hold much more – for example, preferred phone settings, recently dialled numbers, or the equivalent of an address book. (One incidental consequence has been the reduced need to remember phone numbers. I can no longer recall phone numbers of friends, since my phone identifies incoming and outgoing calls by name.)

The introduction of the SIM card was prompted by anxieties over crime – although the ease with which cards can be cloned has erased this advantage. Indeed, the high incidence of cloning again reminds us that any aspect of a new technology of mobility might be exploited for criminal ends. The SIM card also reflected wishes – hopes rather than expectations – that the card, which could identify the purchaser, would spark a boom in mobile commerce. GSM, remember, was imagined as a key infrastructural component of a future European market.

Additionally, the GSM standard provided the option of identifying equipment through an Equipment Identification Number (EIN). This meant the SIM would say who was on the phone and the EIN would say exactly which phone was being used. In the UK, the mobile phone companies chose *not* to activate the EIN capability. In April 2002, the British prime minister, Tony Blair, made an announcement that many commentators considered a hostage to fortune: that the recent rise in crime would be under control by the following September. What seemed like a bold political gamble was, in fact, partly a measured guess that if the mobile companies were persuaded to activate the EIN, then stolen phones could be cut off as soon as they were reported. Crime and mobile phones had become matters of political spin.

Meanwhile, information held by mobile phone operators and even the personal data stored on the SIM card have become detective tools and forensic evidence. In 1993, the hiding place of the cocaine baron Pablo Escobar was found by tracking his mobile radio calls. In October 2002, similar surveillance probably contributed to the arrest of Al Qaida suspect Abu Qatada. Just one other example must serve as an illustration. In July 2001, after four years on the run, the authorities finally caught up in the Philippines with Alfred Sirven, a French businessman wanted as part of the investigations into the Elf corruption scandal. At the moment of arrest, Sirven realised the potentially incriminating evidence he held on his person. James Tasoc, of the National Bureau of Investigation in the Philippines, described what happened next: 'He munched up the chip of his mobile like chewing gum. He broke it with his teeth.'

While the uses of mobile phones for criminal activity and criminal detection are revealing, we should not forget a much more everyday role: many mobile phones are carried as part of the complex strategies that we all develop for dealing with the risks and dangers of modern life. So a phone is bought by parents for a son or daughter who is leaving home as a student, or for a teenager who is starting to stay out late at night, as an act of reassurance. Should they ever be in trouble, they will not lack the means of contact. The diminishing guardianship of family life is replaced by the constant touch of the mobile phone. Again, the fear of car breakdowns in remote places has often been cited by women as a major reason for owning a cell phone. (Notice how the two technologies of mobility interact again.)

Likewise, in countries such as Sweden, the UK or Germany, where most of the population already

Women were targeted in mobile phone advertising in the 1980s. Note how 'peace of mind' is promised if 'you're always in touch'. However, exploiting fears of crime was an old tactic, as the next illustration demonstrates. (By courtesy of BT Archives.)

'Why aren't YOU on the 'phone?', issued by the telephone development association, c. 1905. Evidence that fear of crime has been used to sell telephones since the early days. Thank heavens that 'obliging constables' would save us from 'marauding humanity'! (By courtesy of BT Archives.)

own a mobile phone, further expansion of the market has depended on sales pitches that play on fear. Early adopters of the mobile needed little persuasion, and the young took to the chatty communicability of mobile phones like ducks to water. But selling to an older market, which can be either more techno-phobic or wiser in their purchases, has relied on presenting the mobile phone as a safety technology of last resort. This was why my parents bought one. While anyone who has been saved a walk along the hard shoulder of a motorway at night to find a land line telephone will know the true benefits of the mobile phone as a tool of safety, sometimes the device seems to be merely talismanic. Like a statuette of St Christopher, possession alone seems to ward off danger.

A colleague of mine tells a story of one of his parents who carries a mobile phone in the car. On one occasion a passenger needed to make a phone call and asked to use the mobile. Unsure about its operation, the passenger enquired how to work it. 'Oh, I don't know', came the reply. 'I always have it turned off. It's only for emergencies.'

CHAPTER 16

PHONES ON FILM

Our sense of danger is powerfully shaped by mediated representations of life. Hitch-hiking is now rare in the West, not because hitch-hikers are a danger to drivers, or *vice versa*, but because on film a hitch-hiker tends to be a homicidal maniac. Paedophiles are a menace, but a youngster is far more likely to be run over by someone driving their kids to school (which they do because 'the streets are dangerous') than be abducted. Again, a major cause of the school run is our calculation of risk, in this case to children, based on what we read and see, rather than on what statistics might show.

So how we use one technology of mobility – the car – depends on how it has been portrayed on television, in film or in print. What, then, does the mobile in the media show us? In one straightforward way, it has been a great boon to scriptwriters. A good plot depends on interaction between characters. Communication technologies allow characters who are not in the same room as each other to interact, thereby literally expanding the dramatic range. With a land line telephone, a character is rooted to the spot. But first with cordless phones (which I haven't discussed here) and then mobile phones, characters were set free. The scriptwriters of soaps and sitcoms were perhaps one of the greatest beneficiaries, since soaps and sitcoms depend more than any other televisual genres on conversation and gossip. So an episode of *The Simpsons* or *One Foot in the Grave*, to take just two examples from the USA and the UK, can even be dominated by characters on mobile phones.

Film both created and exploited mobile phones as iconic markers of status. Gordon Gekko, the feral corporate raider played by Michael Douglas in the sharp satire of 1980s excess, *Wall Street* (1987), barked orders down a brick-like cell phone while walking on an Atlantic beach. In Gekko's individualistic, greed-driven world, money never slept – and never stopped moving. Gekko would have no 'dead time'. *Wall Street* repeatedly drew a contrast between this new atomised anti-society and the older traditional society where personal integrity was based on good character and a firm's worth lay in real products, not junk finance. The cell phone was the icon of the new. And fiction in turn shaped fact. Across the ocean, the city workers in

London who ostentatiously waved mobile phones modelled themselves on Gekko, their hero.

So phones on film could symbolise connection between characters, advertise social status, or even stand for the absence of society. But a more intriguing use of the mobile exploited the contradictions – and horror – of being in constant touch. In David Lynch's *Lost Highway* (1997), a creepy atmosphere had already been built up, with the delivery of video tapes made by an intruder into the house of characters played by Patricia Arquette and Bill Pullman. Video tapes are information technologies that shift time. The footage merely showed that the intruder had been in the house at some point in the past. But creepiness turns to full horror at a party, where the uncomfortable Pullman is approached by a white-faced man with more than a whiff of sulphur about him. The white-faced man insists they have met. What's more, he insists that he is in Pullman's home at the same instant as he is there at the party. The proof is a mobile phone call. The mobile is an information technology of instantaneous time. The horror comes from the sudden realisation that the white-faced man is in two places at once, so something is seriously wrong.

There are two ways that mobiles feature in stories of the uncanny, and both reflect contradictory aspects of mobile culture. In TV series such as *The*

X-Files, mobile phones are part of the armoury of the good – in this case the FBI agents Fox Mulder and Dana Scully – against evil. The cell phone here provides horizontal communication between the heroes (Mulder and Scully) who are working outside, and often against, the centralised hierarchical organisation (the FBI).

The same analysis fits *The Matrix* (1999), in which the character played by Keanu Reeves and his allies discover that the world is a simulation created by an authoritarian mechanical regime. They can act in both worlds only because they possess state-of-the-art mobiles. And as entertainment and communications industries converged in the late 1990s, *The Matrix* acted as an advert for the particular phone used, the Nokia 8110i. So, ironically, the mechanical simulation-creating regime won. Indeed, on the day of the film's launch, Heikki Norta, general manager, Marketing Services, Nokia Mobile Phones, Europe and Africa, said:

> *Nokia's mobile phones create the vital link between the dream world and the reality in* The Matrix. *The heroes of the movie could not do their job and save the world without the seamless connectivity provided by Nokia's mobile phones. Even though our everyday tasks and duties may be less important than those of the heroes of*

> The Matrix, *today we can all appreciate the new dimension of life enabled by mobile telephony. As the leading brand in mobile communications, Nokia is proud to see that the makers of* The Matrix *have chosen Nokia's mobile phones to be used in their film.*

In the second way, the creepiness of instantaneous remote communication is exploited. In the *Scream* movies, which knowingly cherry-picked the whole horror genre, the killer was anonymous, remote but also scarily present as soon as the call was made. In *Lost Highway*, the uncanniness stemmed from the impossibility of being in two places at the same time – a short-circuiting of spatial logic. How can someone be both present and not present? Mobile phones give us a powerful sense of co-presence that can be shockingly undermined. I was once chatting to a friend who was walking down the Hackney Road in London. There was a scuffle and then silence. While it was clear what had happened – the phone had been snatched – the shock (for me) lay in sudden helplessness, the realisation that someone who is near is in fact far. Constant touch is illusory.

The mobile phone on film can counter and cause the uncanny. These two modes are part of longer traditions in story-telling. The use of the latest communication technologies to counter ancient evil is nowhere better illustrated than by Bram Stoker's *Dracula* (1897), in which otherwise utterly ordinary modern Europeans can defeat the Count because they possess dictaphones, telegraphs and an efficient postal service. (Recall that Stoker's novel, for good reason, is in epistolary form.) *The Blair Witch Project* (1999) was deliberately set a few years in the past, after cheap video cameras (on which the film's beguiling realism depends) but *before* mobile phones. There was no escape for this second team of ordinary humans.

In *Haunted Media*, Jeffrey Sconce has traced how communications technologies have persistently been associated with the uncanny, from the 'spiritual telegraph' of the 1840s to oppressive other-worlds of fictional cyberspaces in the late twentieth century. So with early radio, catching distant voices by accident across the ether ('distant signal' or 'DX fishing') suggested to many authors a metaphor for the fragile bonds between individuals and the potential for traumatic disconnection. 'Stage and screen at the beginning of the century', notes Sconce, 'saw a number of productions that featured distraught husbands listening helplessly on the phone as intruders in the home attacked their families.'

Later, episodes of *The Twilight Zone* featured uncanny communication by the dead by phone. In

'A Long Distance Call' (1961), a recently deceased grandmother called her grandson on a toy telephone, and in 'Night Call' (1964), a long-dead fiancé contacts a woman via a telephone wire that has fallen on his grave. Horror stems from interrupted mobility – whether it be from confinement to a grave or, traumatically outside fiction, from mobile phones in the twisted wreckage of train crashes or the last conversations on the hijacked airliners of 11 September 2001.

PART IV

REASSEMBLING THE MOBILE AS A GLOBAL SYSTEM

· CHAPTER 17 ·

THE GLOBE IS MADE
BY STANDARDS

In 1950, the major ports of the world swarmed with human activity. The job of a stevedore or long-shoreman, someone who loaded and unloaded ships, was a skilled one: goods could come in a wide variety of shapes and sizes, and the quickest, most efficient, way of moving them had to be worked out. Once dockside, the goods might wait for some time, each minute costing money, before they could be moved to market. Each port had its own system, its own traditions and its own considerable pool of labour, amounting to thousands of dockers.

In the 1960s and 1970s, this picture of the working dock was transformed by containerisation. Goods would be packed in identical steel boxes, making the jobs of lifting on and off ship, and of transporting to and from a port, much simpler (and cheaper). Some historians credit the innovation to the experiments of the US military in the Second World War, in which essential supplies had to be shipped to Europe and across the Pacific in a form that was secure. Others highlight the entrepreneurial spirit of Malcolm McLean, an ex-trucker whose SeaLand Inc. company began shipping goods in containers along the east coast of the USA from 1956. By the 1970s, it was clear that a revolution in the global transport of *material* goods had happened, and that it was underpinned by two fundamental developments.

First, a global technological system for transport went hand in hand with the spread of a single standard. While many agreed that a standard container was a good thing, there had been much debate about what the standard should be. The outcome was a container 8 feet high by 8 feet wide, with lengths either 20, 35 or 40 feet. (Universal agreement on the width and the height were crucial for stacking containers, the length not so. Think about how a sound brick wall can be made with short bricks and long bricks.) If there had been many competing standards in use, then global trade, and with it the forces of globalisation, would have been significantly reduced.

Second, as the standard spread, which it did by a combination of commercial and governmental decisions, the old practices and old infrastructure had to be torn up and replaced. Starting with Port Elizabeth, New Jersey, container ports were built to move the standardised containers onto ships converted for the new boxes or onto trains and trucks. Some old ports – not least London, which had once been the busiest in the world – died. Others, such as Long Beach, USA, were refitted at great cost. The new infrastructure was massively expensive, but was paid for by the savings in transporting goods and in savings of scale. All the world was using one standard. Ninety per cent of the world's trade moves in containers.

A fixed infrastructure, and standards mutually agreed beforehand, facilitated global mobility – in the case of time zones, mobility of pocket watches; in the case of containerisation, mobility of material goods; and in the case of the Internet (where the fixed infrastructure was land lines and the standards were the so-called TCP/IP protocols), mobility of *non-material* goods. (There are profound reasons why the latter two cases of means of moving packets, material and non-material, appeared at the same time.) And, from fairly early in its history, there have been visions of how a fixed worldwide infrastructure of cellular phones would enable the global mobility of communication. We have seen how very different *national* systems of mobile telephony were built, and now we will see what makes a *global* system, and why.

Jorma Niemienen, then president of Mobira,

imagined in 1982 that the mobile world could be built on Nordic lines: 'NMT [Nordic Mobile Telephone] is an example of the direction which must be taken. The ultimate objective must be a worldwide system that permits indefinite communication of mobile people with each other, irrespective of location.'

In Vancouver in 1986, before the first call had ever been made on the second – that is, digital – generation of phones, a gathering of telecommunications planners launched the third. Initially called 'Future Public Land Mobile Telephone System', or FPLMTS ('Unpronounceable in any language', writes Garrard, correctly), 'the initial concept for the third generation was very simple – a pocket-sized mobile telephone that could be used anywhere in the world'.

Third-generation (3G) mobile phones started as a geographical idea, but as the Internet rocketed in the 1990s it provided proof that there was public interest, and a nascent mass market, for mobile on-line services. 3G became more and more a plan for mobile phones that would handle data – Internet-type services, videos, games – as well as voice. The relative successes and failures, respectively, of i-mode in Japan and WAP in Europe and the USA were dressed up as rehearsals, generation two-and-a-half, for the data-rich 3G.

Despite the lessons taught by the cases of competing mobile standards in the USA (or indeed, the success of single standards such as Europe's GSM or McLean's containerisation), third generation mobile technology has splintered into several different standards. So much was at stake – perhaps the biggest telecommunications sector of the twenty-first century – that uniform agreement was perhaps impossible to achieve in the face of divergent commercial interests.

So International Mobile Telecommunications 2000 (IMT-2000, the more friendly name for FPLMTS), became the umbrella for five different standards, employing variations of all three means of packaging up and sending data over mobile networks: TDMA, FDMA and CDMA. Each had different coalitions of backers, reflecting the state of a mobile industry that had already become internationalised after a series of mergers and new operations, led by voracious companies such as Vodafone and Hutchison, and by the Baby Bell companies' attempts to expand away from the restrictive US home markets in the 1990s.

Despite the internationalisation of the mobile sector, an interesting pattern was apparent by 2002: American- and Japanese-based companies were doing better at pushing 3G than their European competitors. In a reversal of the transition from first (analogue) to second (digital) generation cell phones, when the USA lost the lead partly because its first generation was too successful, the success of European second generation systems, particularly GSM, had led to apathy towards 3G.

The first licences for the spectrum space allocated to 3G mobile phones came up for grabs at the very end of the century. With Internet stocks still riding high, and where an auction format was chosen by governments, mobile companies bid against each other, driving the price for spectrum to stratospheric levels. When the bidding was over in the UK, the government received a windfall of £22.47 billion (US$35.4 billion), which was prudently earmarked by the Chancellor of the Exchequer, Gordon Brown, for paying off part of the national debt.

Licences went to the four existing operators – Vodafone Airtouch, One 2 One, BT Cellnet and Orange – and a newcomer, a conglomerate backed by the Hong Kong-based Hutchison Whampoa. In Germany, a bigger potential market than in Britain, the auction raised US$45.6 billion – five times the amount initially expected. France refused to run such a market-driven scheme and preferred to retain central control, offering four licences at fixed prices of US$4.6 billion each. Only two were taken up. Sweden, even more planning-minded, awarded licences at US$10,000 each, plus a cut of profits. The German and UK windfalls were watched jealously in the US, where the practice of local auctions was again followed for the sale of broadband Personal Communication Service (PCS) licences, with the outcome in early 2001 that a mere US$16.86 billion was raised for 422 licences (113 of which went to a joint venture between Vodafone and Verizon Wireless).

But what had caused this American shortfall? By the time of the US auction, the Internet stock bubble was bursting, and the giant bids that seemed necessary to secure important territories and markets in earlier months now seemed decidedly dicey. Indeed, the bids made by companies such as Vodafone were justified by appeal to the strength of their stock market value, and as these slid alongside other telecoms stocks, the expenditure looked more and more untenable. Moreover, 3G cannot operate on the older infrastructure. Entirely new networks of base stations and mobile switching centres needed to be built.

Like the containerisation of ports, the great capital outlay in a gamble on a new standard was to be new infrastructure (the costs are comparable). In 2002, the licences were active but, apart from in places such as the Isle of Man, a third-generation experimental island, very few services were launched until 2003. The success of third generation mobile phones depends on the unknowable willingness of the public to buy them, and without good content – in the form of addictive entertainment or really useful services – a repetition of the WAP debacle was possible. On the other hand, third generation might prove to be like a global i-mode, to the great relief of the world economy.

CHAPTER 18

PERPETUUM MOBILE?

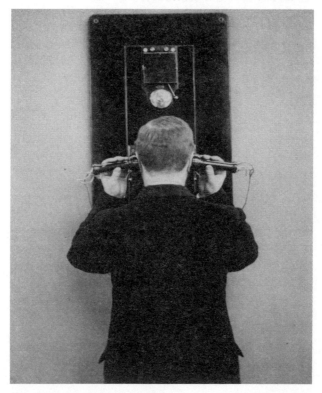

(By courtesy of BT Archives.)

By 2002 there were at least a billion cellular subscribers in the world. The mobile cellular phone has meant different things to different people – for example, a way of rebuilding economies in eastern Europe, an instrument of unification in western Europe, a fashion statement in Finland or Japan, a mundane means of communication in the USA, or an agent of political change in the Philippines. Different nations made different mobiles.

But, through all the contrasting national pictures of cellular use, a common pattern can be glimpsed. There has been a correlation, a sympathetic alignment, between the mobile phone and the horizontal social networks that have grown in the last few decades in comparison with older, more hierarchical, more centralised models of organisation. There is, I feel, a profound sense in which the mobile represents, activates and is activated by these networks in a way that, say, the mainframe computer of the 1950s gave form to the centralised, hierarchical, bureaucratic organisation. What changed between then and now was a social revolution – of which technological change was part and parcel.

The critical attitude towards centralised authority, which emerges in the 1960s, can be seen in examples as diverse as the CB radio fad of the early 1970s, with its alternative jargon and myths of living outside traditional society, or the use of the mobile phone to organise illegal rave parties around London's M25 motorway in the late 1980s and early 1990s. Social revolution produced the youth movement that created demands for a material culture – including cell phones – that met social needs of *distinct* fashions, and *independent* means of communication between friends.

We must be careful not to suggest that technologies of mobility inevitably oppose centralised power. Just over 2,000 years ago, city-states around Italy and the Mediterranean began to feel the force of a new type of power – imperial Rome. These city-states came to accept – and see – themselves as 'Roman', partly because 'Roman' roads gave the social élite an effective means of travel. Anywhere visible from the road became places from which this association of mobility, the technologies enabling mobility and centralised Roman power, could be reinforced. The great triumphal arches, adverts for 'Rome' through which a traveller had to pass, provide one example. As political power became concentrated in the hands first of an oligarchy and senate, and then into the hands of individual emperors, so the road system became ever more important. Technological systems of mobility helped to create the Roman world, and mobility reinforced central hierarchical imperial power.

But try a historical thought experiment. What

would Rome have been like if the technologies of mobility had been truly demotic? If, instead of being built to create and sustain the mobility of a social élite, the roads and road culture of Rome had been for the people? Well, I would suggest, as a parallel to think about, the mobile phone as it became in the last decade of the twentieth century, and the car by mid-century, are just such demotic – and interconnected – technologies of mobility. GSM, 'the most complicated system built by man since the Tower of Babel', was meant to make 'Europeans' as the roads made 'Romans', but it was not dedicated to the maintenance of a privileged élite. (However, in other ways the comparison stands. The power of several mobile-based multi-nationals is becoming imperial in stature, and operator portals, which advertise the riches of such companies and through which we must pass, have more than a passing resemblance to triumphal arches.)

There are fierce tensions between demotic technologies of mobility and centralised power. I've given many throughout this book, but the thought of Rome prompts one more example. In March 2001, the Italian Bishops' Conference, the governing organisation of the Roman Catholic Church in that country, circulated to all parish priests a strongly worded warning not to allow mobile aerial masts on church buildings. The mobile operators had a desperate need for sites for new masts, preferably high up and in the centres of populations, and were willing to pay handsomely. The parish priest, often some distance from the riches of the Vatican, had an expensive church to maintain and falling congregations. While some very worldly concerns fell against the obvious deal – such as legal violations that might endanger churches' tax-exempt status – a greater danger arose: the centrality of Christian symbols would be blurred if the spires also advertised mobile masts. The masts, wrote the Bishops, were 'alien to the sanctity' of churches. And in a world where the Catholic Church, perhaps the prime model of authority and hierarchy, perceived numerous threats, the blurring of symbols of mobility and static power was unconscionable. (Broadcasting, always at ease with hierarchy, was another matter: Vatican Radio was permitted its masts.)

In contrast, the more compromising, more pragmatic or less symbolically minded Church of England signed a deal with Quintel S4, a spin-off of QinetiQ (the commercial arm of Britain's defence research agency) in June 2002 to allow mobile masts on 16,000 churches.

But, if compromises can be made with hierarchies, the demotic mobile has been found to fit within horizontal social networks with greater ease. This was not the discovery of the mobile phone manufacturing or operating companies (although companies such as Nokia can boast a flattened management structure and espouse a pro-innovation, anti-deferential corporate philosophy). Instead, the demotic mobile was the discovery, and reflection, of the users. No one within the industry, for example, expected the extent of the success of SMS. Indeed, even now it is remarkable that a service that works out on average at roughly a penny or cent *a character* was successful at all. But the power of text, like many other aspects of the mobile, was found by the people who used it, not the people who planned it.

The mobile phone in the early twenty-first century is in a moment of transition. The third generation is being launched. There are rival models of wireless communication, some centralised (like satellite system Globalstar), some even more pot-entially demotic (wireless LAN, bluetooth, and a host of other means of passing data from device to device, free of charge). The mobile is in danger on three fronts: technological change might add so many new features, benefiting from greater data-handling capacities, that it might barely act like a phone. Its own flexibility will destroy it. The mobile phone would have been a mere passing stage to another technology. Or the rival means of mobile communication and data handling across *ad hoc* networks will prove more economical, more popular or a better fit with social or political imperatives. The mobile cellular standards, from AMPS to Wide-band-CDMA, will lie abandoned. The immense sums spent on outmoded or unwanted 3G licences will collapse some of the biggest corporations, with significant effects on the rest of the economy. Or, the new mobile will be rejected by a more powerful force, the users.

But there is good reason to suggest that users will continue to love the mobile phone. Back in 1906, the inventor of electronic circuitry, Lee de Forest, made the first radio transmission to an automobile. A press release issued to advertise the achievement expressed de Forest's hope that in the near future 'it will be possible for businessmen, even while automobiling, to be kept in constant touch'.

In the centralised, hierarchical world of Ford, such a dream of constant touch made little sense. While it was technically possible earlier, the cellular mobile phone took off only after the 1960s. By then there had been a sea-change in both politics and technology, one affecting the other, after which, to put it crudely, networks of people prospered and hierarchical styles have suffered.

When I smashed up my mobile phone I wanted

to find out what was in it and what sort of world it made sense to assemble it in. Apart, the debris reflected a fragmented, flexible, atomised world. Put together, the mobile provides a *network*, giving society back a cohesion, of sorts. We've seen that the phone can be assembled in many ways. But only after the great transformation of social attitudes of the 1960s could a world wish to be in constant touch. We live *this* side of that transformation, which is why the mobile will be, if not perpetually in motion, at least moving for some time yet.

Txt Msgs: A Transcript

* In case you are not familiar with the text messaging 'language' commonly used, here is a translation into everyday English of the section entitled 'Txt Msgs' on page 105.

What is culture? Culture is the collection of signs specific to a society. Every technology has a culture of its own. A culture can be hard to understand to outsiders. There is no better illustration of this than text messages.

Text messaging was an accident. No one expected it. When the first text message was sent, in 1993 by Nokia engineering student Riku Pihkonen, the telecommunications companies thought it was not important. SMS – Short Message Service – was not considered a major part of GSM. Like many technologies, the *power* of text – indeed the *power* of the phone – was discovered by users. In the case of text messaging, the users were the young or poor in the West and East.

BIBLIOGRAPHY

Artur Attman, Jan Kuuse, Ulf Olsson, Christian Jacobæus et al, *L M Ericsson 100 Years*, 3 vols, Örebro, 1977 (three-volume company history of Ericsson).

Barry Brown, Nicola Green and Richard Harper (eds), *Wireless World: Social and Interactional Aspects of the Mobile Age*, London: Springer, 2002.

BBC and CNN (for the Congo volcano and mobile phones). For example, www.cnn.com/2002/WORLD/africa/01/20/congo.mobiles/

Business Week. i-mode quotations taken from www.businessweek.com/adsections/sun/heroes/content.html

Cellular News. An excellent resource, at www.cellular-news.com

Council of the European Communities. 87/371/EEC: Council Recommendation of 25 June 1987 on the co-ordinated introduction of public pan-European cellular digital land-based mobile communications in the Community. This and other European documents can be found via www.europa.it

Kristi Essick, 'Guns, money and cell phones', *The Industry*

Standard, 11 June 2001. An informative article on coltan. Also useful was the United States Geological Survey reports available via the USGS website.

Claude S. Fischer, *America Calling: a Social History of the Telephone to 1940*, Berkeley: University of California Press, 1992.

H. N. Gant, *Mobile Radio Telephones: an Introduction to their Use and Operation*, London: Chapman & Hall, 1959.

Garry A. Garrard, *Cellular Communications: Worldwide Market Development*, Boston: Artech House, 1998.

Victoria Harrington and Pat Mayhew, *Mobile Phone Theft*, Home Office Research Study 235, Home Office Research, Development and Statistics Directorate, January 2001.

A. Jagoda and M. de Villepis, *Mobile Communications*, Chichester: John Wiley, 1993. (Published in France, 1991.) A good source for the European factors behind GSM.

James E. Katz and Mark A. Aakhus (eds), *Perpetual Contact: Mobile Communications, Private Talk, Public Performance*, Cambridge: Cambridge University Press, 2002.

Timo Kopomaa, *The City in Your Pocket. Birth of the Mobile Information Society*, Helsinki: Gaudeamus, 2000. Finnish social research, confident that the mobile is reinvigorating public space.

Ray Lawrence, *The Roads of Roman Italy: Mobility and Cultural Change*, London: Routledge, 1999. Roads and an 'alteration in the mentalité of space-time' to produce Romans.

Mobile Communications International. A good source of mobile news and international statistics.

Robert C. Morris, *Between the Lines: A Personal History of the British Public Telephone and Telecommunications Service, 1870–1990*, Just Write Publishing Ltd, 1994.

Public Record Office. HO/255 series has many interesting files relating to early mobile radio in the UK.

Richard Robison and David S. G. Goodman, *The New Rich in Asia: Mobile Phones, McDonalds and Middle-Class Revolution*, London: Routledge, 1996.

Leo G. Sands, *Guide to Mobile Radio*, New York: Greensback Library Inc., 1958.

Jeffrey Sconce, *Haunted Media: Electronic Presence from Telegraphy to Television*, Durham: Duke University Press, 2000. There are a few books in the world that have central ideas so good I wished I had written them. Ritvo's *The Animal Estate* is one, and this is another. It traces an otherwise lost history of media, and historicises (and therefore critiques) modern cybertheory into the bargain.

Dan Steinbock, *The Nokia Revolution: the Story of an Extraordinary Company that Transformed an Industry*, New York: AMACOM, 2001.

Alan Stone, *How America Got On-Line: Politics, Markets, and the Revolution in Telecommunications*, Armonk: M. E. Sharpe, 1997.

Peter Young, *Person to Person: the International Impact of the Telephone*, Cambridge: Granta Editions, 1991.

Alan D. Wallis, *Wheel Estate: the Rise and Decline of Mobile Homes*, Oxford: Oxford University Press, 1991.

Other science titles available from
Icon Books:

Dawkins vs. Gould

Kim Sterelny

'Book of the Month' – *Focus* magazine

'Slim and readable ... the aficionado of evolutionary
theory and the intense debate it engenders would do well
to read it.' *Nature*

'A deft little book ... its insights are both useful and fun'
The Australian

Science has seen its fair share of punch-ups over the
years, but one debate, in the field of biology, has become
notorious for its intensity. Over the last twenty years,
Richard Dawkins and Stephen Jay Gould have engaged in
a savage battle over evolution that shows no sign of
waning.

Kim Sterelny moves beyond caricature to expose the
real differences between the conceptions of evolution of
these two leading scientists. He shows that the conflict
extends beyond evolution to their very beliefs in science
itself; and, in Gould's case, to domains in which science
plays no role at all.

ISBN 1-84046-249-3   Paperback   £5.99

The Discovery of the Germ

John Waller

From Hippocrates to Louis Pasteur, the medical profession relied on almost wholly mistaken ideas as to the cause of infectious illness. Bleeding, induced vomiting and mysterious nostrums remained staple remedies. Surgeons, often wearing butcher's aprons caked in surgical detritus, blithely spread infection from patient to patient.

Then came the germ revolution: after two decades of scientific virtuosity, outstanding feats of intellectual courage and bitter personal rivalries, doctors at last realised that infectious diseases are caused by microscopic organisms.

Perhaps the greatest single advance in the history of medical thought, the discovery of the germ led directly to safe surgery, large-scale vaccination programmes, dramatic improvements in hygiene and sanitation, and the pasteurisation of dairy products. Above all, it set the stage for the brilliant emergence of antibiotic medicine to which so many of us now owe our lives.

In this book, John Waller provides a gripping insight into twenty years in the history of medicine that profoundly changed the way we view disease.

ISBN 1-84046-373-2   Hardback   £9.99

An Entertainment for Angels

Patricia Fara

'A concise, lively account.' Jenny Uglow, author of *The Lunar Men* (2002)

'Neat and stylish ... Fara's account of Benjamin Franklin's circle of friends and colleagues brings them squabbling, eureka-ing to life.' *The Guardian*

'Vividly captures the ferment created by the new science of the Enlightenment ... Fara deftly shows how new knowledge emerged from a rich mix of improved technology, medical quackery, Continental theorising, religious doubt and scientific rivalry.' *New Scientist*

'Combines telling anecdote with wise commentary ... presents us with numerous tasty and well-presented historical morsels.' *Times Higher Education Supplement*

Electricity was the scientific fashion of the Enlightenment, 'an Entertainment for Angels, rather than for Men'. Patricia Fara tells the engrossing tale of the strange birth of electrical science – from a high-society party trick to a symbol of man's emerging dominance over nature.

ISBN 1-84046-348-1   Hardback   £9.99

Eureka!

Andrew Gregory

'Marvel as Andrew Gregory explains how the Greeks destroyed myths and gods in favour of a rule-based cosmos ... A readable, pocket-sized primer and a worthwhile present for anyone who needs to fill in the gaps in their knowledge.' *New Scientist*

Eureka! shows that science began with the Greeks. Disciplines as diverse as medicine, biology, engineering, mathematics and cosmology all have their roots in ancient Greece. Plato, Aristotle, Pythagoras, Archimedes and Hippocrates were amongst its stars – master architects all of modern, as well as ancient, science. But what lay behind this colossal eruption of scientific activity?

Free from intellectual and religious dogma, the Greeks rejected explanation in terms of myths and capricious gods, and, in distinguishing between the natural and the supernatural, they were the first to discover nature. New theories began to be developed and tested, leading to a rapid increase in the sophistication of knowledge, and ultimately to an awareness of the distinction between science and technology.

Andrew Gregory unravels the genesis of science in this fascinating exploration of the origins of Western civilisation and our desire for a rational, legitimating system of the universe.

ISBN 1-84046-289-2   Hardback   £9.99

How Far is Up?

John & Mary Gribbin

How far is it to the edge of the Universe? Less than eighty years ago astronomers began to realise that the Milky Way galaxy in which we live is just one island in an immense ocean of space.

Award-winning authors John and Mary Gribbin tell the story of how the cosmic distance scale was measured, the personalities involved and the increasingly sophisticated instruments they used. Astronomers can now study light from objects so distant that it has taken ten billion years on its journey across space to us, travelling all the time at a speed of 300,00 kilometres per second: that's how far up is!

ISBN 1-84046-439-9   Hardback   £9.99

Knowledge is Power

John Henry

Francis Bacon, the renowned English statesman and man of letters, is a leading figure in the history of science. Yet he never made a major discovery, provided a lasting explanation of any physical phenomena or revealed any hidden laws of nature. How then can he rank alongside the likes of Isaac Newton – one of the finest scientists of them all?

Bacon was the first major thinker to describe how science should be done, and to explain why it should be done that way. Against the tide of his times, he rejected the gathering of scientific knowledge for its own sake. Instead, he saw the bounty of science in terms of practical benefit to mankind, and its advance as a means to improve the daily lives of his contemporaries. But foremost, and thus making by far his greatest contribution, Bacon promoted the use of experimentation, coming to outline and define the rigorous procedures of the 'scientific method' that today forms the very bedrock of modern scientific progress.

In this fascinating and accessible book, John Henry gives a dramatic account of the background to Bacon's innovations and the sometimes unconventional sources for his ideas. He explains how magic, civil service bureaucracy and the belief in a forthcoming apocalypse came together in the creation of Bacon's legacy, why he was so concerned to revolutionise the attitude to scientific knowledge – and why his ideas for reform still resonate today.

ISBN 1-84046-356-2   Hardback   £9.99

Latitude & the Magnetic Earth
Stephen Pumfrey

'A chunky read with much more to it than first meets the eye. [Stephen Pumfrey] marshals his scientific and philosophical themes impressively while adding flesh to the hitherto enigmatic Gilbert.' *New Scientist*

'This bijou volume is most valuable for its insights … It is as much for his method as for his conclusions that we should remember this great Elizabethan.' *TLS*

William Gilbert (1544–1603) was royal physician to Queen Elizabeth I and the most distinguished man of science to emerge from her reign. He is the inventor of the term 'electricity', the father of electrical studies, the creator of modern magnetic science and – most famously – the discoverer of the Earth's magnetic nature. Yet, incredibly, he is largely unknown.

Gilbert's close contact with the elite mariners of Elizabethan London enabled him to learn of the magnetic compass and of the strange behaviour of its magnetised needle – a phenomenon known as the magnetic 'dip'. Using a pioneering experimental method, he came to realise that the Earth is a giant magnet; a great body imbued with a 'magnetic soul' that drove it forward in its Copernican orbit. In this golden age of circumnavigations of the globe and of the founding of new colonies, he was the first to use magnetism to determine the latitude of a ship at sea. Alongside these discoveries, Gilbert's writings – some even proposing to solve the problem of longitude – challenged the scientific orthodoxy of his day, and boldly led the battle to establish our modern ideas of terrestrial magnetism.

Lively and accessible, *Latitude & the Magnetic Earth* –
the first new exploration of Gilbert for forty years –
brings the story up to date, leaving the reader with a vivid
feel not only for the conflicts surrounding Gilbert's
discoveries and his scientific legacy, but for the man
himself.

ISBN 1-84046-290-6   Hardback   £9.99

Lovelock & Gaia

Jon Turney

Gaia is a theory that has revolutionised how we view this
'pale blue dot': all living things are part of one great
organism, and life as a whole shapes the planetary
environment.

Opponents dubbed it mere metaphor or myth, that
was either irrefutable or unnecessary, or argued that it is
impossible – after Darwin – for life to affect environment
on a global scale in any way which could be fruitfully
coupled with natural selection.

It is not clear where it will lead, but its impact com-
pares with the greatest of scientific revolutions.

ISBN 1-84046-458-5   Hardback   £9.99

The Manhattan Project

Jeff Hughes

Established in 1942 at the height of the Second World War, the Manhattan Project was a dramatic quest to beat the Nazis to a deadly goal: the atomic bomb. At Los Alamos and several other sites, American, British, Canadian and refugee European scientists, together with engineers, technicians and many other workers, laboured to design and build nuclear weapons. Their efforts produced 'Little Boy' and 'Fat Man', the bombs that ultimately destroyed Hiroshima and Nagasaki in August 1945.

A vast and secret 'state within a state', the Manhattan Project cost $2 billion. It catapulted scientists – particularly nuclear scientists – to positions of intellectual prestige and political influence. State funds flowed for science as never before, and led to the creation of huge new research institutes, especially large particle accelerators designed to explore the properties of matter – like that at CERN, near Geneva. With their huge experiments, complex organisation and lavish funding, these institutes represented a new form of scientific organisation: 'Big Science'.

Yet, from the large astronomical telescopes of the nineteenth century to the factory-like laboratories of the 1930s, 'Big Science' has a social and scientific history that long pre-dates the advent of the atom bomb. Arguing that the Manhattan Project both drew on and accelerated a trend already well underway, Jeff Hughes offers a lively reinterpretation of the key elements in the history and mythology of twentieth-century science.

ISBN 1-84046-376-7   Hardback   £9.99

Moving Heaven and Earth

John Henry

'It's an ideal introduction to Copernicus and helps us understand his key role as one of the founders of the modern world.' *Morning Star*

When Nicolaus Copernicus claimed that the Earth was not stationary at the centre of the universe but circled the Sun, he brought about a total revolution in the sciences and consternation in the Church – a twin upheaval that would eventually lead to the trial of Galileo before the Inquisition in Rome. His astronomical theory demanded a new physics to explain motion and force, a new theory of space, and a completely new conception of the nature of our universe.

But that wasn't all. The theory that moved heaven and earth also showed for the first time that a common-sense view of things isn't necessarily correct, and that mathematics – no matter how abstract it might seem – can and does reveal the *true* nature of the material world. No other single innovation could have had the same far-reaching consequences in sixteenth-century society, where pure knowledge was thought to rest only in surviving fragments of Ancient wisdom.

Copernicus sowed the seed from which science has grown to be a dominant aspect of modern culture, fundamental in shaping our understanding of the workings of the cosmos. In this book, John Henry not only explains how these changes followed upon Copernicus's theory, but also reveals why, in the first place, Copernicus was led to such a seemingly outrageous and implausible idea as a swiftly moving Earth.

ISBN 1-84046-251-5   Paperback   £5.99

Perfect Copy

Nicholas Agar

Cloning represents some of the most exciting – and some of the most morally complex – science of our time.

In 1997 Ian Wilmut and his team announced that they had done the impossible: they had cloned a mammal from an adult cell. This breakthrough prompted immediate calls for the new technology to be used on humans. Italian fertility specialist Severino Antinori hopes to use cloning to give infertile couples the opportunity to at last become parents. Cloning may also solve, once and for all, the problem of rejection that bedevils transplant surgery. Perhaps it even holds the secret of eternal life.

But plans to clone humans have triggered an international storm of protest. Scientists, including Wilmut, politicians from left and right, and theologians from almost all religions find the idea not just unsavoury, but abhorrent.

In this book, Nicholas Agar provides a uniquely accessible exploration of this highly controversial issue. Starting with the biology, and building up the scientific background step by step, *Perfect Copy* provides the perfect guide to the moral labyrinth that surrounds the cloning debate.

ISBN 1-84046-380-5   Paperback   £7.99

Turing and the Universal Machine

Jon Agar

'An excellent treatment … highly readable.' *New Scientist*

The history of the computer is entwined with that of the modern world and most famously with the life of one man, Alan Turing. A machine unlike any other, this 'electronic brain' is of apparently *universal* application; yet paradoxically, given its almost infinite scope, it can only follow instructions. How did this device, which first appeared a mere 50 years ago, come to structure and dominate our lives so totally?

Turing, widely hailed as the man instrumental in breaking the Nazi Enigma code, is also regarded as the father of the modern computer. In this book, Jon Agar tells the fascinating history of the appearance of the universal machine: from the work of Charles Babbage in the 1820s and 30s, and the data-sorting nightmare of the 1890 American Census, to Turing's formulation of a 'computing machine' designed to solve an infamous mathematical problem of his day, and his later explorations into Artificial Intelligence. Spurred on by the imperatives of the Second World War, the first commercial electronic computer was built in 1951 and nicknamed the 'Blue Pig'. Yet Turing did not live long enough to celebrate its success. A victim of Cold War paranoia, his prosecution for homosexuality led to a severing of his connections with the British secret service, and shortly after to his suspected suicide in 1954.

Setting events in a rich historical context, *Turing and the Universal Machine* makes the development of the computer readily understandable but no less remarkable.

ISBN 1-84046-250-7   Paperback   £5.99

Watt's Perfect Engine

Ben Marsden

James Watt is synonymous with the steam engine, that Promethean symbol of the Industrial Revolution. But what motivated Watt to re-invent steam? What convinced him that the stunningly simple idea of giving Thomas Newcomen's ubiquitous fire-engine a 'separate condenser' could work in practice, banish waste – and achieve perfection? And how did Watt's perfect engine become the progenitor of progress, and its problems, in nineteenth-century Britain?

The astonishing success of steam meant taking a tiny classroom toy and thinking big, re-equipping industry with a new factory power. That meant cashing in on connections and re-deploying instrument makers' skills. It meant conspiring with enlightened philosophers, like chemist Joseph Black, ensconced in Glasgow's ancient College, and the hard-nosed entrepreneurs, like bucklemaker Matthew Boulton, who consorted for the sake of commerce at Birmingham's Lunar Society.

This is a tale of science and technology in tandem, of factory show-spaces and international espionage, of bankruptcy and braindrains, lobbying and legislation, and patents and pirates. It is a story of boiling kettles and leaking cylinders, mechanisms perfected, monopolies defended, and competitors trounced. And this is a book about another kind of perfection: taking the man James Watt, warts and all, and making an icon fit for an age of invention.

ISBN 1-84046-361-9   Hardback   £9.99

Postmodernism and Big Science

Edited by Richard Appignanesi

The last hundred years, eclipsing even the Enlightenment in breadth of scientific innovation, have witnessed science emerge as the definitive paradigm with which to explain the natural world.

Technological advances have allowed scientists to penetrate far beyond the everyday world and speculate widely. When and how was the Universe created? Have humans descended from apes? Can we measure time? Philosophy, it seems, has lost metaphysics to Big Science.

Postmodernism and Big Science examines five figures instrumental in this sea change: Darwin, Dawkins, Einstein, Hawking and Kuhn. This book is a fascinating journey through the debates of our time, and a tonic to the idea that their resolution may be imminent.

ISBN 1-84046-351-1   Paperback   £8.99

In case of difficulty in obtaining any Icon title through normal channels, books can be purchased through BOOKPOST.

Tel: + 44 1624 836000
Fax: + 44 1624 837033
E-mail: bookshop@enterprise.net
www.bookpost.co.uk

Please quote 'Ref: Faber' when placing your order.

If you require further assistance, please contact:
info@iconbooks.co.uk